Nonlinear acoustics through problems and examples

Oleg V. Rudenko
Sergey N. Gurbatov
Claes M. Hedberg

2009

Order this book online at www.trafford.com
or email orders@trafford.com

Most Trafford titles are also available at major online book retailers.

Printed in Victoria, BC, Canada.

ISBN: 978-1-4269-0544-5 (sc)

ISBN: 978-1-4269-0545-2 (dj)

ISBN: 978-1-4269-0546-9 (e-book)

Our mission is to efficiently provide the world's finest, most comprehensive book publishing service, enabling every author to experience success. To find out how to publish your book, your way, and have it available worldwide, visit us online at www.trafford.com

Trafford rev. 2/9/2010

 www.trafford.com

North America & international
toll-free: 1 888 232 4444 (USA & Canada)
phone: 250 383 6864 ♦ fax: 812 355 4082

Contents

iv

Foreword

Nonlinear acoustics is based mainly on results obtained more than 30 years ago. These results have appeared in monographs [1-5] and were discussed in special problem-oriented programs and lectures on the theory of waves [6]. The next stage of the nonlinear acoustics development is associated with the onset of the wide use of its concepts and methods in applied fields [7-9]. The latter required numerical solutions of nonlinear equations describing one-dimensional waves [13], and beams [14]. Waves in inhomogeneous media [15] and randomly modulated perturbations [13-15] were studied. Modern studies in fundamental and applied nonlinear acoustics are described in new books and reviews (see, for example [16-21]).

Nowadays, nonlinear acoustics is considered to be a well-developed branch of science and technology providing results which can be advantageously employed by specialists working in different fields. It may be useful for undergraduate and postgraduate students, as well as the specialists in neighbouring scientific directions, to have a textbook which would allow them, within a comparatively short period of time, to grasp the fundamentals of nonlinear acoustics and to become active users of the methods developed here concerning simplifications, calculations and numerical estimates.

We believe that a sequence of problems properly ordered from a logical viewpoint can be regarded among the most efficient ways of grasping new ideas. The problems dealt with in this book come in groups. As a rule, the first problem in a group is aimed at careful examination of an important theoretical aspect and is followed by a comprehensive solution. The subsequent problems serve the purpose of mastering the methods of calculations and making estimates. The

1

2

simplest ones have nothing but an answer, whereas more complicated problems have both an answer and an explanation. The most difficult tasks are being privileged by having a solution. When a group consists of problems of the same type a solution is only provided for the first one, while the others in the group are meant to be considered in a similar manner. We have tried to adhere to this layout where it is possible.

This book of problems stems from the programs offered to the students of the Moscow and Gorky[1] University Acoustics Chairs as well as to the students of the radiophysics Department of the Moscow State University Physics Faculty. Many new problems were suggested by the authors in the course of preparing the manuscript. Sections 1,2,3 are joint efforts of the authors. Problems 1.4, 1.7, 1.9, 1.19-1.24, 2.2, 2.6, 2.10, 2.15, 2.17, 3.2, 3.4, 3.6-3.13 were suggested by S.N. Gurbatov in their initial version while the others are due to O.V. Rudenko; section 4 is based on the suggestions of O.V. Rudenko and section 5 is a contribution from S.N. Gurbatov. The whole body of the problems has been perfected for a long time by means of cross-checking and editorial activity. The authors are very greatful to V.A. Khokhlova and V.A. Gusev who took the burden of scientific editing of the manuscript. It was according to their suggestions that certain solutions were modified. Since courses of nonlinear acoustics are being delivered in various universities and technical colleges in Russia, Sweden, USA, Germany, France, Great Britain, China, Japan and many other countries, we dare hope that this book will be reader-friendly and add to the preparation of specialists in the field. We are ready to discuss every remark or suggestion forwarded to us by those interested and aimed at improving and supplementing our future editions.

O.V. Rudenko, S.N. Gurbatov, and C.M. Hedberg

The foreword and chapters 1-5 were translated from Russian by D.G. Sorokin.

---

[1]Since October 11, 1990 again Nizhny Novgorod

INTRODUCTION

For the subsequent solving of nonlinear problems, it is first necessary to give some information about linear acoustic waves.

The system of equations describing the fluid or gas with account for shear η and bulk ζ viscosities consists of the Navier-Stokes equation of motion

$$\rho\left[\frac{\partial \vec{u}}{\partial t} + (\vec{u}\nabla)\,\vec{u}\right] = -\nabla p + \eta\,\Delta\vec{u} + \left(\zeta + \frac{\eta}{3}\right)\operatorname{grad}\operatorname{div}\vec{u}\ , \qquad (\text{I.1})$$

the continuity equation

$$\frac{\partial\rho}{\partial t} + \operatorname{div}(\rho\,\vec{u}) = 0\ , \qquad (\text{I.2})$$

and the equation of state which for acoustic waves has the form of a Poisson adiabat

$$p = p\,(\rho) = p_0\left(\frac{\rho}{\rho_0}\right)^{\gamma}\ . \qquad (\text{I.3})$$

The Eulerian approach to continuous media is used in the formulation of the system (I.1)-(I.3). In the context of this approach all variables – the pressure p, the density ρ and the velocity u – are functions of coordinates of an immovable reference frame, and of time t. Often, another approach is used, namely the Lagrangian one which is helpful for solving one-dimensional problems (see Problems 1.1-1.5). The Lagrangian description of the continuous medium uses coordinates of liquid particles measured at a definite initial moment of time, instead of the coordinates of an immovable reference frame. The Lagrangian approach is the basic one in the theory of elasticity of solids and is mainly concerned with the positional relationship of different internal sections of solids which is responsible for the appearance of stress and strain fields. At the same time, the rigid displacements of solid bodies are less interesting.

Let the undisturbed state of a fluid be $\rho = \rho_0$, $p = p_0$, and $\vec{u} = 0$. Disturbances caused by the wave are denoted by the primed letters and are put into the system (I.1)-(I.3)

$$p = p_0 + p'\,, \ \ \rho = \rho_0 + \rho'\ . \qquad (\text{I.4})$$

The disturbances are considered to be small:

$$\frac{p'}{p_0} \sim \frac{\rho'}{\rho_0} \sim \frac{|\vec{u}|}{c_0} \sim \mu \ll 1 \ . \tag{I.5}$$

Here μ is a small parameter. A power series expansion in μ will be used to simplify the differential equations. The equilibrium sound velocity is

$$c_0 = \sqrt{\left(\frac{\partial p}{\partial \rho}\right)_{\rho=\rho_0}} = \sqrt{\gamma \frac{p_0}{\rho_0}} \ . \tag{I.6}$$

The ratio of the particle velocity to the sound velocity is known as the acoustic Mach number M. One can see from equation (I.5) that μ is of the same order as the acoustic Mach number. Substituting (I.4) into the system (I.1)-(I.3), one can reduce it to the form

$$\rho_0 \frac{\partial \vec{u}}{\partial t} + \nabla p' - \eta \Delta \vec{u} - \left(\zeta + \eta/3\right) \operatorname{grad} \operatorname{div} \vec{u} =$$
$$= -\rho' \frac{\partial \vec{u}}{\partial t} - (\rho_0 + \rho')\left(\vec{u}\nabla\right)\vec{u} \tag{I.7}$$

$$\frac{\partial \rho'}{\partial t} + \rho_0 \operatorname{div} \vec{u} = -\operatorname{div}\left(\rho' \vec{u}\right) \ , \tag{I.8}$$

$$p' = c_0^2 \rho' + \frac{1}{2}\left(\frac{\partial^2 p}{\partial \rho^2}\right)\rho'^2 + \ldots \equiv A\frac{\rho'}{\rho_0} + \frac{1}{2}B\left(\frac{\rho'}{\rho_0}\right)^2 + \ldots \ . \tag{I.9}$$

The linear terms are collected in left-hand-sides of equations (I.7) and (I.8), and the nonlinear terms containing powers and products of the variables describing disturbances are collected in the right-hand-sides. In keeping only the linear terms, the system (I.7)-(I.9) is reduced to

$$\rho_0\frac{\partial \vec{u}}{\partial t} + \nabla p' - \eta \Delta \vec{u} - \left(\zeta + \frac{\eta}{3}\right)\operatorname{grad}\operatorname{div}\vec{u} = 0, \tag{I.10}$$

$$\frac{\partial \rho'}{\partial t} + \rho_0 \operatorname{div}\vec{u} = 0 \ , \tag{I.11}$$

$$p' = c_0^2 \rho' \ . \tag{I.12}$$

The deviation of density in equation (I.11) can be excluded by using equation (I.12). As a result, a pair of equations is derived which consists of the following one:

$$\frac{\partial p'}{\partial t} + c_0^2 \rho_0 \operatorname{div} \vec{u} = 0 \tag{I.13}$$

and equation (I.10), which connects the two variables p' and $\vec{u}$.

In accordance with the Helmholtz theorem, an arbitrary vector field (in our case the velocity field) can be represented as the sum of potential and vortex components:

$$\vec{u} = \vec{u}_l + \vec{u}_t, \qquad \vec{u}_l = \nabla \varphi, \qquad \vec{u}_t = \operatorname{rot} \vec{A} \ . \tag{I.14}$$

The function φ is a so-called scalar or acoustic potential, and the function $\vec{A}$ is the vector potential. Substituting (I.14) into equation (I.13), and taking into account the identity $\operatorname{div} \operatorname{rot} \vec{A} = 0$, we derive

$$\frac{\partial p'}{\partial t} + c_0^2 \rho_0 \operatorname{div} \vec{u}_l = 0 \ : \tag{I.15}$$

One can see that the pressure variation (as well as the variation of density (I.12)) is caused only by the potential (the acoustic) component of the velocity field. The vortex component describes the shear motion where the medium behaves as an incompressible liquid. It follows from equation (I.12) that the shear motion is described by the diffusion equation:

$$\frac{\partial \vec{u}_t}{\partial t} = \frac{\eta}{\rho_0} \Delta \vec{u}_t \ . \tag{I.16}$$

To describe the potential natural mode of the medium, the following equation can be derived from equation (I.10):

$$\rho_0 \frac{\partial \vec{u}_l}{\partial t} + \nabla p' = b \Delta \vec{u}_l, \qquad b \equiv \left(\zeta + \frac{4}{3} \eta \right) \ . \tag{I.17}$$

Apply the gradient operator to equation (I.15), then differentiate equation (I.17) with respect to time and subtract this from the former.

There after one can eliminate the pressure variation and find a single equation for the potential component $\vec{u}_l$:

$$\frac{\partial^2 \vec{u}_l}{\partial t^2} - c_0^2 \Delta \, \vec{u}_l = \frac{b}{\rho_0} \frac{\partial}{\partial t} \Delta \vec{u}_l \; . \tag{I.18}$$

Let us now consider a plane wave where all parameters depend on one of the coordinates (for example, x) and time (t). From the potentiality condition it follows that the velocity vector has only one nonzero component, which is its projection on the x-axis. This component will be denoted hereafter as u. Consequently, the acoustic wave is a longitudinal wave for which the particles of the medium vibrate along the direction of its propagation. For a plane wave equation (I.18) takes form (see Problem 3.1)

$$\frac{\partial^2 u}{\partial t^2} - c_0^2 \frac{\partial^2 u}{\partial x^2} = \frac{b}{\rho_0} \frac{\partial^3 u}{\partial t \, \partial x^2} \; . \tag{I.19}$$

For example, let a monochromatic wave propagate along -axis:

$$u\,(x,\,t) = A \exp\left(-i\omega t + ikx\right) \; , \tag{I.20}$$

where k and ω are the wave number and the frequency. After substitution of (I.20) in (I.19), the dispersion law is obtained:

$$k^2 = \frac{\omega^2}{c_0^2}\left(1 - i\frac{b\omega}{c_0^2 \rho_0}\right)^{-1}, \quad k = \pm\frac{\omega}{c_0}\left(1 - i\frac{b\omega}{c_0^2 \rho_0}\right)^{-1/2} \; . \tag{I.21}$$

The minus sign in the second formula in (I.21) corresponds to a wave traveling in the positive direction of the x-axis, and the minus sign to a wave traveling in the negative direction. The dimensionless combination of parameters inside the brackets in (I.21), which contains the effective viscosity b, is assumed to be small:

$$\frac{b\omega}{c_0^2 \rho_0} \ll 1 \; . \tag{I.22}$$

Then the dispersion relation (I.21) takes form

$$k = \frac{\omega}{c_0} + i\frac{b\omega^2}{2c_0^3 \rho_0} \; . \tag{I.23}$$

Substituting (I.23) in (I.20), we derive

$$u\left(x,\,t\right) = A\exp\left(-\frac{b\omega^2}{2c_0^3\rho_0}x\right)\exp\left(-i\omega\,t + i\,\frac{\omega}{c_0}x\right)\ .\qquad(\text{I}.24)$$

One can see, that the wave amplitude decreases according to the law

$$A\left(x\right) = A\exp\left(-\alpha\,x\right),\qquad \alpha \equiv \frac{b\omega^2}{2c_0^3\rho_0}\ .\qquad(\text{I}.25)$$

The combination of parameters α in (I.25) is the absorption coefficient of the wave. If absorption over the distance of one wave length λ is small, or in other words,

$$\alpha\lambda = \frac{b\omega^2}{2c_0^3\rho_0}\frac{2\pi\,c_0}{\omega} = \pi\frac{b\omega}{c_0^2\rho_0} \ll 1\ ,\qquad(\text{I}.26)$$

then the assumption (I.22) is true. Really, a wave can be *defined* as a process which can transport energy over distances much greater than the wave length λ. The most typical and important case is precisely the one corresponding to inequality (I.26).

If the viscosity (wave absorption) is negligible, one can derive from equations (I.12),(I.15),(I.17) the following linear relations for acoustic disturbances traveling in positive direction along x:

$$\frac{p'}{c_0^2\rho_0} = \frac{\rho'}{\rho_0} = \frac{u}{c_0}\ .\qquad(\text{I}.27)$$

For the wave propagating in negative direction the particle velocity in formula (I.27) contains a minus sign. Consequently, any disturbance (acoustic pressure, acoustic density or particle velocity) is described by the same wave equation. A simpler equation can be derived if we restrict the problem and consider only the wave propagating in one direction, say, in positive direction of the x-axis. Replace in the dispersion relation (I.23) the wave number and frequency by differential operators according to (I.20):

$$k \to -i\frac{\partial}{\partial x}\ ,\qquad \omega \to i\frac{\partial}{\partial t}\ .\qquad(\text{I}.28)$$

A second order equation is obtained (instead of the third-order equation (I.19))

$$\frac{\partial u}{\partial x} + \frac{1}{c_0}\frac{\partial u}{\partial t} = \delta\,\frac{\partial^2 u}{\partial t^2}\ , \qquad \delta \equiv \frac{b}{2c_0^3\rho_0}\ . \tag{I.29}$$

It is more convenient to replace (x,t) with the new set of variables

$$x,\ \tau = t - x/c_0\ . \tag{I.30}$$

By the introduction of a time which moves together with the wave at the sound velocity (I.30), we remove the fast process (wave propagation) and watch the slow distortion (evolution) of the wave caused by weak viscosity. In the accompanying coordinate system (I.30) equation (I.29) takes the form of the common diffusion equation (see (3.2)):

$$\frac{\partial u}{\partial x} = \delta\,\frac{\partial^2 u}{\partial \tau^2}\ . \tag{I.31}$$

A generalized form of this equation with account for nonlinearity is the Burgers equation (3.8).

Using the approach described above we can get another important linear equation which governs the evolution of spatially bounded wave beams. The dispersion relation for the three-dimensional wave equation for media without absorption is

$$\frac{\omega^2}{c_0^2} = k^2 \equiv k_x^2 + k_y^2 + k_z^2\ . \tag{I.32}$$

Here k_x, k_y, and k_z are projections of the wave vector on the axes.

Let the wave beam propagate along the x-axis while weakly diverging or converging. It means that the projections k_y and k_z are small in comparison with k_x and the following approximate relation is valid

$$\frac{\omega}{c_0} = k_x\sqrt{1 + \frac{k_y^2 + k_z^2}{k_x^2}} \approx k_x + \frac{k_y^2 + k_z^2}{2k_x}\ ,$$

$$k_x\left(\omega - c_0 k_x\right) = \frac{c_0}{2}\left(k_y^2 + k_z^2\right). \tag{I.33}$$

By replacing the wave vector and frequency components with operators in the second formula in (I.33) (by analogy with the formula (I.28))

$$k_x \rightarrow -i\frac{\partial}{\partial x}, \; k_y \rightarrow -i\frac{\partial}{\partial y}, \; k_z \rightarrow -i\frac{\partial}{\partial z}, \; \omega \rightarrow i\frac{\partial}{\partial t}, \qquad (I.34)$$

we derive

$$\frac{\partial}{\partial x}\left(\frac{\partial u}{\partial t} + c_0\frac{\partial u}{\partial x}\right) = \frac{c_0}{2}\left(\frac{\partial^2 u}{\partial y^2} + \frac{\partial^2 u}{\partial z^2}\right). \qquad (I.35)$$

Passing on to the accompanying coordinate system (I.30), one obtains the evolution equation (see (4.20)):

$$\frac{\partial^2 u}{\partial x\,\partial \tau} = \frac{c_0}{2}\left(\frac{\partial^2 u}{\partial y^2} + \frac{\partial^2 u}{\partial z^2}\right). \qquad (I.36)$$

A generalization of this equation for the nonlinear case leads to the Khokhlov-Zabolotskaya equation (4.23).

Chapter 1

SIMPLE WAVES

Problem 1.1

Show that a set of equations of hydrodynamics in Lagrangian variables for a one-dimensional plane motion has a solution in the form of simple Riemann waves. Reduce this set to a single nonlinear equation for the variable $\xi(x, t)$ - which is the displacement of the medium particles from their initial position x.

Solution The reference equations of hydrodynamics in Lagrangian representation have the form [3],[4]

$$\rho_0 \frac{\partial^2 \xi}{\partial t^2} = -\frac{\partial p}{\partial x} , \quad \rho_0 = \rho(1 + \frac{\partial \xi}{\partial x}) , \quad p = p(\rho) = p_0(\frac{\rho}{\rho_0})^\gamma . \quad (1.1)$$

The first equation is a generalization of the second Newton's law with respect to a continuous medium. The second one (the continuity equation) is the law of conservation of mass written down in differential form. The third one is the equation of state given in the form of a Poisson adiabat for fast processes of compression and rarefaction (as compared with thermodiffusion) accompanying the sound propagation.

With a simple wave is understood a wave formulation (a nonlinear one, speaking generally) in which all the variables can be expressed through one another with the help of some algebraic relations. If, however, the relations of variables contain integrals or derivatives, the wave will not be considered as simple; from a physical viewpoint this means the appearance of dispersion, i.e. the dependence of even a small perturbation behaviour on its spectral composition. As the latter two equations of (1.1) are formulated as

$$\rho = \rho_0/(1 + \frac{\partial \xi}{\partial x}) , \quad p = p(\rho) = p(\rho_0/1 + \frac{\partial \xi}{\partial x})) , \quad (1.2)$$

the density and pressure are expressed as functions of only one variable $\partial\xi/\partial x$. This implies that the set (1.1) has a solution in the form of simple waves.

Consider the equation of state (1.1) in the form of an adiabat. Then (1.2) yields $p = p_0(1 + \partial\xi/\partial z)^{-\gamma}$. Substituting this relationship into the right-hand side of the first equation in (1.1) results in the nonlinear Earnshaw equation

$$\frac{\partial^2\xi}{\partial t^2} = c_0^2\frac{\partial^2\xi/\partial x^2}{(1 + \partial\xi/\partial x)^{\gamma+1}}, \tag{1.3}$$

where $c_0 = (\gamma p_0/\rho_0)^{1/2}$ is the equilibrium sound velocity. The equation (1.3) has a nonlinearity of a general type and can formally be used to describe strong disturbances. However, it is required, that the denominator in (1.3) will not vanish, i.e. $(\partial\xi/\partial x \neq -1)$. In nonlinear acoustics we are dealing with weakly nonlinear waves for which $|\partial\xi/\partial x| \ll 1$.

Problem 1.2

Considering weak nonlinearity, simplify the Earnshaw equation (1.3) and retain only the two principal nonlinear terms.

Solution Make use of the approximate relationship

$$(1 + \frac{\partial\xi}{\partial x})^{-(\gamma+1)} \approx 1 - (\gamma + 1)\frac{\partial\xi}{\partial x} + \frac{1}{2}(\gamma + 1)(\gamma + 2)\left(\frac{\partial\xi}{\partial x}\right)^2 . \tag{1.4}$$

Substitute the expansion (1.4) into the right-hand side of Earnshaw equation (1.3). Rewrite it in the form:

$$\frac{\partial^2\xi}{\partial x^2} - \frac{1}{c_0^2}\frac{\partial^2\xi}{\partial t^2} = (\gamma + 1)\frac{\partial\xi}{\partial x}\frac{\partial^2\xi}{\partial x^2} - \frac{1}{2}(\gamma + 1)(\gamma + 2)\left(\frac{\partial\xi}{\partial x}\right)^2\frac{\partial^2\xi}{\partial x^2} . \tag{1.5}$$

The left-hand side of (1.5) corresponds to a typical linear wave equation. The right-hand side is resulting from the general type of nonlinearity expansion in terms of power nonlinearities which contains quadratic and cubic nonlinear terms.

Problem 1.3

A nonlinear medium occupies a half-space $x > 0$ and at its boundary $x = 0$ a harmonic signal $\xi = A\sin\omega t$ with angular frequency ω is prescribed. Find out which frequencies that can appear during wave propagation in a medium due to quadratic and cubic nonlinearities, by analysing the equation (1.5) using the method of successive approximations.

Solution By considering nonlinear effects to be weak, the right-hand side in equation (1.5) can be neglected in the first approximation. A solution to the linear wave equation in the form of a wave travelling in the positive direction of the axis will be expressed as

$$\xi^{(1)}(x,t) = A\sin\omega(t - x/c_0) \ . \tag{1.6}$$

In order to find a solution to the second approximation one has to substitute (1.6) into the right-hand side of the nonlinear equation which will consequently take the form

$$F = \frac{1}{2}(\gamma + 1)A^2\left(\frac{\omega}{c_0}\right)^3 \sin 2\omega\tau +$$

$$\frac{1}{8}(\gamma + 1)(\gamma + 2)A^3\left(\frac{\omega}{c_0}\right)^4(\sin 3\omega\tau + \sin\omega\tau) \ , \tag{1.7}$$

where $\tau = t - x/c_0$ is the time in the "accompanying" coordinate system which travels with the wave at sound velocity c_0. The equation of the second approximation with the right-hand side (1.7) can be written as

$$\frac{\partial^2\xi^{(2)}}{\partial x^2} - \frac{1}{c_0^2}\frac{\partial^2\xi^{(2)}}{\partial t^2} = F(t - x/c_0) \ . \tag{1.8}$$

It is evident that F has the meaning of an "exciting force" in an inhomogeneous wave equation (1.8). It generates new waves at the frequencies of the second harmonic 2ω (quadratic nonlinear effect) and the third harmonic 3ω (cubic nonlinear effect). Apart from this, the cubic nonlinearity gives additional contribution to the wave at the fundamental frequency ω (the self-action effect).

Problem 1.4

Indicate which frequencies that emerge in a quadratic nonlinear medium (in the first approximation) provided a biharmonic signal $\xi = A_1 \sin \omega_1 t + A_2 \sin \omega_2 t$ is given as input. Consider in particular the limiting case $\omega_1 \to \omega_2$.

Solution In analogy with problem 1.3 it can be readily shown that the second harmonics $2\omega_1$, $2\omega_2$ of the reference frequency waves are generated in the medium, as well as perturbations at sum $\omega_1 + \omega_2$ and difference $\omega_1 - \omega_2$ frequencies. For $\omega_1 \to \omega_2$, the second harmonic alone will be generated inasmuch as the difference frequency excitation efficiency tends to zero (see also Problem 1.7).

Problem 1.5

Simplify equation (1.5) by using the method of slowly varying profile retaining only a quadratic nonlinear term.

Solution The method of slowly changing profile allows a significant simplification of the nonlinear partial differential equations describing the process of propagation of intense waves. Simplified equations will naturally lend themselves to easier solutions. The concept of the method is as follows. Given the absence of nonlinear terms, a solution to equation (1.5) will look like a sum of two travelling waves of arbitrary form: $\xi = \Phi(t - x/c_0) + \Psi(t + x/c_0)$. The wave with profile $\Phi(\tau)$ propagates in the positive direction of the x-axis whereas the wave Ψ travels in the negative direction. We take an interest in the positive one. When a weak nonlinearity is observed and the right-hand side of the equation differs from zero, the wave form will cease to be constant - it will deform as it propagates. Provided the nonlinearity is weak, the wave profile is changing slowly. i.e. along with the "fast" dependence of the function Φ on $\tau = t - x/c_0$, a slow dependence of Φ on x must develop

$$\xi = \Phi(\tau = t - x/c_0, x_1 = \mu x) . \tag{1.9}$$

Here $\mu \ll 1$ is the small parameter of the problem conforming to the smallness of nonlinear terms in equation (1.5) as compared with linear ones

$$\mu \sim (\gamma + 1)\left|\frac{\partial \xi}{\partial x} \cdot \frac{\partial^2 \xi}{\partial x^2}\right| / \left|\frac{\partial^2 \xi}{\partial x^2}\right| \sim (\gamma + 1)\left|\frac{\partial \xi}{\partial x}\right| \ll 1 . \qquad (1.10)$$

The fact that $\left|\frac{\partial \xi}{\partial x}\right| \ll 1$ has been used in passing from the Earnshaw equation (1.3) to the simplified equation (1.5). If we assume, in particular, that the displacement variation is in accordance with the harmonic law $\xi = A \sin \omega(t - x/c_0)$ the smallness condition will acquire the form

$$\mu \sim (\gamma + 1)A\omega/c_0 = (\gamma + 1)2\pi A/\lambda \ll 1 . \qquad (1.11)$$

This means that the particle displacement amplitude A must be small compared to the wave length λ. In other words, the ratio of the oscillation speed amplitude u_0 to the sound speed c_0, u_0/c_0 (acoustic Mach number), has to be a small quantity. Therefore, the small parameter of the problem will be the acoustic Mach number $M = u_0/c_0$.

Let us in equation (1.5) pass from x and t to new variables x_1 and τ in line with the assumption (1.9). Calculate the derivatives

$$\frac{\partial^2 \xi}{\partial t^2} = \frac{\partial^2 \xi}{\partial \tau^2} , \quad \frac{\partial \xi}{\partial x} = -\frac{1}{c_0}\frac{\partial \xi}{\partial \tau} + \mu\frac{\partial \xi}{\partial x_1} ,$$

$$\frac{\partial^2 \xi}{\partial x^2} = \frac{1}{c_0^2}\frac{\partial^2 \xi}{\partial \tau^2} - \frac{2\mu}{c_0} \cdot \frac{\partial^2 \xi}{\partial x_1 \partial \tau} + \mu^2\frac{\partial^2 \xi}{\partial x_1^2} . \qquad (1.12)$$

Substituting (1.12) into equation (1.5) and neglecting all the terms of the order of μ^2, μ^3 and of the higher infinitesimal orders (one needs to take into consideration that the right-hand side of the equation is small compared with the left-hand one) yields

$$\frac{\partial u}{\partial x} = \frac{\epsilon}{c_0^2}u\frac{\partial u}{\partial \tau} , \qquad (1.13)$$

where $u = \partial \xi/\partial \tau = \partial \xi/\partial t$ is the oscillation speed of the medium particles, and $\epsilon = (\gamma + 1)/2$ is a parameter of acoustic nonlinearity.

The equation (1.13) in nonlinear acoustics is called an equation of simple waves. It is noteworthy that this is an equation of the first order rather than of the second order as the original one; therefore, the problem has been simplified to a large extent. Derivation of equation (1.13) from the hydrodynamics equations in the Eulerian representation can be found elsewhere, for example in [4],[6].

The value of the nonlinear parameter can be expressed as $\epsilon = (\gamma + 1)/2 = 1 + B/2A$, where $\gamma = c_p/c_v$ is for gases and A , B are the series expansion factors when expanding the pressure fluctuations in terms of the density increment $p' = A \cdot (\rho'/\rho_0) + (B/2) \cdot (\rho'/\rho_0)^2 + \ldots$:

$^\circ$ C	0	20	40	60	80	100
ϵ	3.1	3.5	3.7	3.8	4.0	4.1

Table 1.1: *Nonlinearity parameter ϵ for distilled water*

For example, sea water at $S = 3.5\%$ and $t = 20^\circ$ C has $\epsilon = 3.65$. For water with steam- and gas-bubbles it depends on the bubble size, the bubble concentration and the frequency, and can be as high as $5 \cdot 10^3$. A few other types of liquids have the values shown in Table 2.1.

Medium	Value of ϵ for 20 $^\circ$ C
Methanol	5.8
Ethanol	6.3
Acetone	5.6
Glycerol	5.4
Transformer oil	4.2
Petrol	6.6

Table 1.2: *Nonlinearity parameter ϵ for different liquids*

Problem 1.6

At the boundary of a nonlinear medium, $x = 0$, the particle velocity follows the law $u(x = 0, \tau = t) = u_0 \sin(\omega t)$. Solving the simple wave equation (1.13) by the method of successive approximations, define the law of the second harmonic amplitude variation with increase in distance x.

Solution From the equation for simple waves we obtain equations in the first and second approximations

$$\frac{\partial u^{(1)}}{\partial x} = 0 \ , \qquad \frac{\partial u^{(2)}}{\partial x} = \frac{\epsilon}{2c_0^2} \cdot \frac{\partial}{\partial \tau} (u^{(1)})^2 \ . \qquad (1.14)$$

The solution of the first approximation $u^{(1)} = u_0 \sin \omega \tau$ is substituted into the second equation (1.14). Integrating it, provided that $u^{(2)}(0, \tau) = 0$ (no second harmonic at the medium boundary), we find that

$$u^{(2)} = (\epsilon/2c_0^2) \, \omega u_0^2 x \cdot \sin 2\omega \tau \ . \qquad (1.15)$$

It is clear that the second harmonic amplitude increases linearly with the x-coordinate. The distance

$$x = x_S = c_0^2/(\epsilon \omega u_0) = \lambda/(2\pi \epsilon M) \ , \qquad (1.16)$$

at which the second harmonic amplitude formally attains $1/2$ of the first harmonic amplitude is defined as a characteristic nonlinear length, or the discontinuity formation distance. In reality, however, solution (1.15) holds true for distances $x \ll x_S$ since, under conditions of considerable energy transfer from the first harmonic to the second one, the solutions obtained by the method of successive approximations are inexact. Formula (1.15) implies that for acoustic signals, which always have a small Mach number ($M \ll 1$), the nonlinear length $x_S \gg \lambda$. In other words, to have its profile and spectrum markedly distorted, a wave has to cover a distance equal to many wave lengths λ. This is exactly what we call the "slowness" of profile variation over the scales on the order of λ (see Problem 1.5).

Problem 1.7

At a boundary $x = 0$ the perturbation is a sum of harmonic signals $u(0, t) = u_1 \sin \omega_1 t + u_2 \sin \omega_2 t$. Solving the simple wave equation (1.13) with the aid of successive approximations, find the amplitudes u_+ and u_- of combination harmonics $\omega_1 + \omega_2$ and $\omega_1 - \omega_2$. Compare the generation efficiency for the sum and difference frequencies.

Answer In analogy with Problem 1.6 we find that

$$u_{\pm} = \frac{\epsilon}{2c_0^2} u_1 u_2 (\omega_1 \pm \omega_2) x \,, \qquad \frac{u_-}{u_+} = \frac{|\omega_1 - \omega_2|}{\omega_1 + \omega_2} < 1 \,. \qquad (1.17)$$

Problem 1.8

Show that an exact solution to the simple wave equation (1.13), for the arbitrary shape perturbation $u(x = 0, t) = \Phi(t)$ at the nonlinear medium boundary, is given by the implicit function

$$u(x, t) = \Phi(\tau + \epsilon u x / c_0^2) \,. \qquad (1.18)$$

Obtain formula (1.18) by the method of characteristics known from the theory of quasilinear partial differential equations of the first order.

Solution Differentiating (1.18) one finds

$$\frac{\partial u}{\partial x} = \frac{(\epsilon/c_0^2) u \Phi'}{1 - (\epsilon/c_0^2) x \Phi'} \,, \qquad \frac{\partial u}{\partial \tau} = \frac{\Phi'}{1 - (\epsilon/c_0^2) x \Phi'} \,, \qquad (1.19)$$

where the prime denotes a derivative with respect to a complete argument of function Φ. Substituting (1.19) into the simple wave equation (1.13) yields identity. Solution by the method of characteristics is described in Problem 2.2. It is interesting to find out the way nonlinear effects are hidden in the implicit relation (1.18). Expressing (1.18) in a series for small x we obtain

$$u \approx \Phi(\tau) + \Phi'(\tau) \frac{\epsilon}{c_0^2} u x + \ldots \approx \Phi(\tau) + \frac{\epsilon}{c_0^2} x \Phi'(\tau) \Phi(\tau) \,. \qquad (1.20)$$

It is seen that the second term is quadratic in the function Φ, i.e. it describes quadratic nonlinear effects. The terms that follow correspond to nonlinearities of higher degrees.

Problem 1.9

By using the implicit solution (1.18) to the simple wave equation (1.13), consider the evolution of the linear profile reference perturbation

$$u(x = 0, t) = \Phi(t) = \gamma(t - t_*) . \tag{1.21}$$

Discuss the cases $\gamma > 0$ and $\gamma < 0$.

Solution Substituting (1.21) into the general formula (1.18) yields

$$u(x, \tau) = \gamma(\tau + \epsilon u(x, \tau)x/c_0^2 - t_*) , \tag{1.22}$$

and, therefore,

$$u = \frac{\gamma(\tau - t_*)}{1 - \epsilon\gamma x/c_0^2} . \tag{1.23}$$

Thus, for any distance x the profile remains linear in τ, the only change being its slope angle with respect to the τ-axis. When the slope is positive ($\gamma > 0$) the solution is valid over the finite interval $x < c_0^2/\epsilon\gamma$ until the profile becomes vertical. For a negative slope ($\gamma < 0$) the profile grows still less steep with x and for $x \gg c_0^2/\epsilon|\gamma|$ the information on the initial slope is actually lost:

$$u \approx -(c_0^2/\epsilon x)(\tau - t_*) . \tag{1.24}$$

Expression (1.23) is the simplest but still rather important solution of the simple wave equation. It can be employed to describe evolution of arbitrary profile portions which are close to straight lines.

Problem 1.10

Perform a graphic analysis of the nonlinear distortions of a period of a reference harmonic signal $u(0, \tau) = u_0 \sin \omega\tau$. Make use of the implicit solution (1.18) to the simple wave equation rewriting it as an explicit function of the variable $\tau(x, u)$

$$\tau = \Phi^{-1}(u) - (\epsilon/c_0^2)ux , \tag{1.25}$$

where $\Phi^{-1}$is the inverse function of Φ.

Solution For $x = 0$, formula (1.25) yields $\tau(0, u) = \Phi^{-1}(u)$, the curve that corresponds to the reference wave profile. The curve corresponding to the nonlinearly distorted profile is obtained on the plane (u, τ) through graphically adding the reference curve and the straight line $-(\epsilon/c_0^2)ux$ - the slope of which grows with x.

For a harmonic signal, with $x = 0$, the solution (1.18) can be formulated as

$$\frac{u}{u_0} = \sin(\omega\tau + \frac{\epsilon}{c_0^2}\omega ux) = \sin(\omega\tau + z\frac{u}{u_0}) , \qquad (1.26)$$

where

$$z = (\epsilon/c_0^2)\omega u_0 x = x/x_S \qquad (1.27)$$

is the distance measured in number of discontinuity formation lengths (1.16). Formula (1.25) for the signal under consideration will take the form

$$\omega\tau = \arcsin\frac{u}{u_0} - z\frac{u}{u_0} . \qquad (1.28)$$

Putting u/u_0 on the abscissas axis, $\omega\tau$ along the ordinate, and carrying out the constructions above one will obtain the pattern depicted in Figure 1.1. It is evident that as the distance covered by the wave increase, the leading front (facing the motion direction) grows steeper, whereas the slope of the trailing edge decreases. A similar picture is observed for sea surface waves as they approache a shore. At a distance of $z = 1$ ($x = x_S$) the leading edge becomes vertical and a discontinuity or a shock front emerges. For $x > 1$ the profile becomes multivalued, i.e. the solution in the form of a simple wave (1.26) is not valid at distances $x > x_S$.

Problem 1.11

Using a solution of the joining type (1.23), consider the evolution of a single triangular pulse with duration $2T$. For $x = 0$ the profile is approximated by the piecewise linear function

$$\frac{u}{u_0} = 0 \ \ (\tau < 0, \ \tau > 2T) , \qquad \frac{u}{u_0} = \frac{\tau}{T} \ \ (0 < \tau < T) ,$$

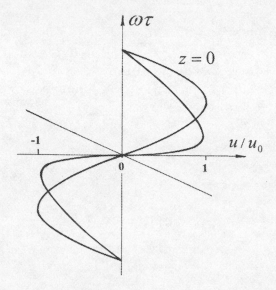

Figure 1.1: **Problem 1.10** *Graphical analysis of the process of distortion of one period of a Riemann (simple) wave during its propagation in a nonlinear medium. The shape of the initial signal is sinusoidal in time and its evolution is described by the solution (1.28).*

$$\frac{u}{u_0} = 2 - \frac{\tau}{T} \quad (T < \tau < 2T) \, . \tag{1.29}$$

Consider the cases $u_0 > 0$ and $u_0 < 0$. Carry out an analysis using the graphic procedure (see Problem 1.10).

Answer

$$u = 0 \, , \quad (\tau < 0 \, , \, \tau > 2T)$$

$$\frac{u}{u_0} = \frac{\tau}{T}(1 - \frac{\epsilon u_0}{c_0^2 T}x)^{-1} \, , \quad (0 < \tau < T - \frac{\epsilon u_0}{c_0^2}x)$$

$$\frac{u}{u_0} = \frac{2T - \tau}{T}(1 + \frac{\epsilon u_0}{c_0^2 T}x)^{-1} \, , \quad (T - \frac{\epsilon u_0}{c_0^2}x < \tau < 2T) \tag{1.30}$$

The solution in the form of a simple wave holds up to the distance

$x = x_S = c_0^2 T/(\epsilon u_0)$ where the leading edge becomes "vertical". The profile distortion process for $u_0 > 0$ is depicted in Figure 1.2.

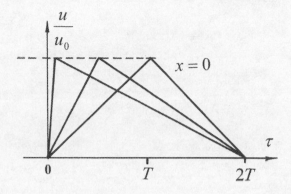

Figure 1.2: *Evolution of the profile of an initial single triangular pulse at its propagation in a nonlinear medium.*

Problem 1.12

Find the spectrum of a simple wave in a nonlinear medium if the wave at the input is given as $u_0(0, \tau) = u_0 \Phi(\omega \tau)$ where Φ is a function which is periodic in its argument with period $T = 2\pi$.

Solution One needs to calculate the coefficients C_n used in the Fourier series expansion of the implicit solution (1.18) to the simple wave equation (1.13)

$$\frac{u}{u_0} = \Phi(\omega\tau + z\frac{u}{u_0}) = \sum_{n=-\infty}^{+\infty} C_n(z)\exp(in\omega\tau) . \qquad (1.31)$$

Here $z = (\epsilon/c_0^2)\omega u_0 x$ is a dimensionless distance. The expansion coefficients are equal to

$$C_n(z) = \frac{1}{2\pi}\int_T \Phi(\omega\tau + z\frac{u}{u_0})\exp(-in\omega\tau)\,d(\omega\tau) . \qquad (1.32)$$

The first integration by parts provides

$$C_n(z) = \frac{1}{2\pi i n} \int_T \exp(-in\omega\tau) d\Phi = \frac{1}{2\pi i n} \int_T \exp(-in[\xi - z\Phi(\xi)]) d\Phi(\xi) \ .$$
$$(1.33)$$

In formula (1.33) we passed to the variable $\xi = \omega\tau + zu/u_0$, which leads to $\omega\tau = \xi - z\Phi(\xi)$, and now the integral contains an explicit function of ξ. Performing the second integration by parts we come to

$$C_n(z) = -\frac{i}{2\pi n z} \int_{-\pi}^{\pi} [\exp(inz\Phi(\xi)) - 1] \exp(-in\xi) \, d\xi \ . \qquad (1.34)$$

As $z \to 0$ we can expand the exponent under the integral (1.34) into a series and obtain an evident result of the linear approximation

$$C_n(z) = \frac{1}{2\pi} \int_{-\pi}^{\pi} \Phi(\xi) \exp(-in\xi) \, d\xi = C_n(z = 0) = \text{constant} \ , \quad (1.35)$$

i.e. no interaction of harmonics occurs and the parameters in the medium are equal to their reference values.

Problem 1.13

Making use of the answer in the previous problem (formula (1.34)) find the harmonics' amplitude dependencies on distance $z = x/x_S$ provided a harmonic signal $u(0, \tau) = u_0 \sin\omega\tau$ is prescribed. Find the power laws of the amplitude growth for $z \ll 1$.

Solution Let us exploit a mathematical identity from the theory of Bessel functions [22],

$$\exp(iz\cos\phi) = \sum_{k=-\infty}^{+\infty} i^k J_k(z) \exp(ik\phi) \ . \qquad (1.36)$$

With the help of identity (1.36), the exponent under integral (1.34) can be expressed as

$$\exp(inz\sin\xi) = \sum_{k=-\infty}^{+\infty} J_k(nz) \exp(ik\xi) \ . \qquad (1.37)$$

After this, the integral can be readily calculated as

$$C_n(z) = -\frac{i}{nz} \sum_{k=-\infty}^{+\infty} J_k(nz)\delta_{nk} = -\frac{i}{nz} J_n(nz) . \qquad (1.38)$$

Defining the real Fourier series expansion coefficients A_n for $\cos n\omega\tau$ and B_n for $\sin n\omega\tau$:

$$A_n(z) = C_n + C_n^* = 0 ,$$

$$B_n(z) = i(C_n - C_n^*) = 2J_n(nz)/(nz) , \qquad (1.39)$$

we obtain the well-known Bessel-Fubini solution

$$\frac{u}{u_0} = \sin(\omega\tau + z\frac{u}{u_0}) = \sum_{n=1}^{\infty} \frac{2J_n(nz)}{nz} \sin(n\omega\tau) . \qquad (1.40)$$

Dependencies of the harmonic amplitudes B_n on the distance are visualized in Figure 1.3. Using the first terms of the Bessel function expansion into the series

$$J_n(x) \approx (x/2)^n/n! , \qquad (1.41)$$

we can write

$$B_n \approx (nz/2)^{n-1}/n! . \qquad (1.42)$$

More exact power approximations and tabulated numerical values of harmonic amplitudes are given in [10].

Problem 1.14

Calculate the variation with distance of the difference frequency wave amplitude provided a biharmonic signal $u/u_0 = \sin\omega_1 t + \sin\omega_2 t$ is specified at the input $x = 0$. It is assumed that $\omega_1 = (N+1)\omega$ and $\omega_2 = N\omega$, where $N > 1$ is an integer.

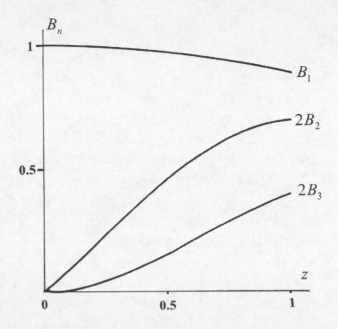

Figure 1.3: **Problem 1.13** *Distance-dependent amplitudes of the first, second and third harmonics of an initially sinusoidal wave. The spectral content is governed by the Bessel-Fubini formula (1.40).*

Solution Inasmuch as the difference frequency is equal to $\omega_1 - \omega_2 = \omega$, we are only interested in the coefficient $C_1(z)$ in (1.34). Using the relationship (1.37) for a biharmonic signal we obtain

$$e^{iz \sin(N+1)\xi + iz \sin N\xi} = \sum_{k=-\infty}^{\infty} J_k(z) e^{ik(N+1)\xi} \sum_{m=-\infty}^{\infty} J_m(z) e^{imN\xi} . \quad (1.43)$$

Substituting this expression into (1.34) shows that the integral differs from zero for $k(N+1) + mN = 1$ alone. This is possible only for the values $k = 1$, $m = -1$ and, therefore, we find that

$$C_1 = iJ_1^2(z)/z , \quad A_1 = 0 , \quad B_1 = 2J_1^2(z)/z . \quad (1.44)$$

Problem 1.15

For $N \gg 1$, within the conditions outlined for the previous problem, define the difference frequency wave amplitude at a distance equal to the discontinuity formation length $z = z_S$. Make a comparison with the result due to the method of successive approximations (see Problem 1.6) and find out the way this amplitude depends on the frequency ratio between ω and ω_1.

Answer $z_S \approx 1/(2N)$, $B_1 \approx z_S/2 \approx \omega/(4\omega_1)$

Problem 1.16

Calculate the behaviour of the amplitudes of the low frequency harmonics which are introduced in a nonlinear medium as a result of the self-demodulation of the reference amplitude-modulated signal $u/u_0 = (1 - m\cos\omega t) \cdot \sin(N\omega t)$, where $N \gg 1$ is an integer, and m is a modulation strength.

Answer In analogy with Problem 1.14 a result is achieved in the form of a series containing the Bessel function products. The main terms of these series have the form

$$C_1 = -\frac{2i}{z} J_1(z) J_0(\frac{m}{2}z) J_1(\frac{m}{2}z) , \qquad C_2 = \frac{i}{2z} J_0(2z) J_1^2(mz) . \quad (1.45)$$

For small z we obtain expressions corresponding to the solution by the method of successive approximations $B_1 \approx mz/2$, $B_2 \approx -m^2z/4$.

Problem 1.17

Consider interaction of a powerful low-frequency wave with a weak high-frequency signal $u(0,t)/u_0 = \sin\omega t + m\sin N\omega t$, $(m \ll 1$, integer $N \gg 1)$. In what manner does the weak signal amplitude vary in space ?

Solution Taking the smallness of m into consideration, formula (1.34) yields

$$C_n(z) \approx \frac{m}{2\pi} \int_{-\pi}^{\pi} \sin(N\xi)\, e^{iNz\sin\xi} e^{-iN\xi}\, d\xi \approx -i\frac{m}{2} J_0(Nz) \ . \quad (1.46)$$

Hence $A_N = 0$ and $B_N = m J_0(Nz)$. The solution (1.46) holds within the range before the discontinuity formation (for $z < 1$). As $N \gg 1$ the Bessel function argument in (1.46) can be a large quantity; in this case the weak signal amplitude will oscillate in space to eventually die down gradually. This effect of nonlinear suppression of a high-frequency signal due to an intense low-frequency disturbance (e.g. noise) is of interest in some practical situations.

Problem 1.18

Using the method of successive approximations, analyze the degenerate parametric interaction in simple waves. For a reference perturbation $u/u_0 = \sin 2\omega t + m \sin(\omega t + \phi)$ with $m \ll 1$, define under which phase shift ϕ the weak signal is amplified and under which ϕ it is suppressed.

Solution Parametric amplification of weak signals in the field of an intense pump wave is considered to be an important nonlinear effect from a practical viewpoint. If the pump frequency is 2ω while that of a signal is ω, the process is defined as degenerate; it is sensitive to the phase shift ϕ between these two waves. In problem 1.6 the equations (1.14) of the first and the second approximation are formulated. Recall that $u^{(1)}$ is a reference perturbation in which $\tau = t - x/c_0$ is placed instead of t; $u^{(2)}$ is the solution of the second approximation to be found. Holding the Fourier-component on the right-hand side of an equation for $u^{(2)}$ at the signal frequency ω yields

$$\frac{\partial u^{(2)}}{\partial x} = -\frac{\epsilon}{2c_0^2} m\omega u_0^2 \sin(\omega\tau - \phi) \ . \quad (1.47)$$

The solution at the signal frequency has the form

$$\frac{1}{u_0}(u^{(1)} + u^{(2)}) = m\sin(\omega\tau + \phi). - \frac{m}{2} z \sin(\omega\tau - \phi) \ . \quad (1.48)$$

Hence it can be seen that the signal amplitude for $z \ll 1$ behaves like

$$2|C_1(z)| = m[\cos^2 \phi(1 - z/2)^2 + \sin^2 \phi(1 + z/2)^2]^{1/2} \approx$$

$$m(1 - z\cos(2\phi))^{1/2} . \tag{1.49}$$

Provided the phase shift ϕ varies from 0 to π, amplification is observed within the range $\pi/4 \leq \phi \leq 3\pi/4$ such that the most efficient amplification of the signal occurs for $\phi = \pi/2$. Within the ranges $0 \leq \phi \leq \pi/4$ and $3\pi/4 \leq \phi \leq \pi$ the signal is suppressed, mostly at $\phi = 0$ and at $\phi = \pi$. It seems useful to solve this problem with the spectral representation (1.34) of the simple wave equation solution (in analogy with Problems 1.14 and 1.17) rather than resorting to the method of successive approximations.

Problem 1.19

Find a Fourier-transform of a simple wave $u(x, \tau)$

$$C(x, \omega) = \frac{1}{2\pi} \int_{-\infty}^{\infty} u(x, \tau) e^{-i\omega\tau} \, d\tau \tag{1.50}$$

assuming that the perturbation vanishes as $\tau \to \pm\infty$.

Solution Employing the general solution (1.18) for a Fourier-transform of a simple wave, one obtains

$$C(x, \omega) = \frac{1}{2\pi} \int_{-\infty}^{\infty} \Phi(\tau + \frac{\epsilon}{c_0^2} ux) e^{-i\omega\tau} \, d\tau . \tag{1.51}$$

Like in the similar problem 1.12, in which a periodic signal was dealt with, one needs to pass over to a new variable $\xi = \tau + (\epsilon/c_0^2)x \cdot u$. Then $\tau = \xi - (\epsilon/c_0^2)x\Phi(\xi)$, and for (1.51) we have the explicit expression

$$C = \frac{1}{2\pi} \int_{-\infty}^{\infty} \Phi(\xi)(1 - \frac{\epsilon}{c_0^2}x\frac{d\Phi(\xi)}{d\xi}) e^{-i\omega(\xi - (\epsilon/c_0^2)x\Phi(\xi))} \, d\xi . \tag{1.52}$$

A even more convenient form can be achieved by integrating (1.52) twice by parts and taking into account that $\Phi(\pm\infty) = 0$:

$$C(x, \omega) = \frac{1}{2\pi i(\epsilon/c_0^2)\omega x} \int_{-\infty}^{\infty} [e^{i(\epsilon/c_0^2)\omega x\Phi(\xi)} - 1] e^{-i\omega\xi} \, d\xi . \tag{1.53}$$

As $x \to 0$ formula (1.53) yields

$$C(x,\omega) = \frac{1}{2\pi} \int_{-\infty}^{\infty} \Phi(\xi)\, e^{-i\omega\xi}\, d\xi = C_0(\omega) \ , \qquad (1.54)$$

being the Fourier-transform of the reference perturbation.

Problem 1.20

Proceeding from solution (1.53) to the previous problem, find a universal behaviour of the Fourier-transform within the low-frequency range. Demonstrate that if $n > 1$ and $C_0(\omega) \sim \omega^n$ as $\omega \to 0$, a universal asymptotic spectrum behaviour is formed due to nonlinear interactions between the spectral components in the low-frequency range ($\omega \to 0$).

Solution Within the low-frequency range the exponent in the solution can be expanded into a series. Restricting ourselves to the terms which are quadratic in ω we come to the expression

$$C \approx \frac{1}{2\pi} \int_{-\infty}^{\infty} \Phi(\xi)\, e^{-i\omega\xi}\, d\xi + \frac{i}{4\pi} x \frac{\epsilon}{c_0^2}\omega \int_{-\infty}^{\infty} \Phi^2(\xi)\, e^{-i\omega\xi}\, d\xi \ . \qquad (1.55)$$

Taking the Fourier-transform property into account

$$\frac{1}{2\pi} \int_{-\infty}^{\infty} \Phi^2(\xi)\, e^{-i\omega\xi}\, d\xi = \int_{-\infty}^{\infty} C_0(\Omega) C_0(\omega - \Omega)\, d\Omega \ , \qquad (1.56)$$

we reduce (1.55) to the form

$$C(x,\omega) = C_0(\omega) + \frac{i}{2} \cdot \frac{\epsilon}{c_0^2}\omega x \int_{-\infty}^{\infty} C_0(\Omega) C_0(\omega - \Omega)\, d\Omega \ . \qquad (1.57)$$

It follows that for the reference spectra of the type $C_0(\omega) \sim \omega^n$, $n > 1$ the low-frequency wave spectrum in a nonlinear medium is described by the universal expression

$$|C(x,\omega)| \approx \frac{\epsilon}{2c_0^2}\Gamma\omega x \ , \qquad \Gamma = \frac{1}{2\pi} \int_{-\infty}^{\infty} \Phi^2(\xi)\, d\xi = \int_{-\infty}^{\infty} |C_0(\Omega)|^2\, d\Omega \ . \qquad (1.58)$$

Problem 1.21

Proceeding from expression (1.53) for the simple wave spectrum, find the Fourier-transform of a signal conforming to the sinusoidal perturbation at the input $\Phi = u_0 \sin \omega_0 t$.

Answer Using the Bessel functions relation (1.36) and the property of the δ-function

$$\delta(\omega) = \frac{1}{2\pi} \int_{-\infty}^{+\infty} e^{-i\omega t} \, dt \; , \tag{1.59}$$

one can write

$$C(x, \omega) = -iu_0 \sum_{k=-\infty}^{\infty} \frac{J_k(kz)}{kz} \delta(\omega - k\omega_0) \; , \tag{1.60}$$

where $z = (\epsilon/c_0^2)\omega_0 u_0 x = x/x_S$. After the inverse Fourier-transform of formula (1.60) we arrive at the Bessel-Fubini solution (1.40).

Problem 1.22

Find the spectrum components $C_{int}(x, \omega)$ resulting from the interaction of an intense pump wave $u_1(t)$ with a weak signal $u_2(t)$.

$$u(x = 0, t) = \Phi(t) = u_1(t) + u_2(t)$$

$$u_1(t) = u_0 \sin \omega_0 t \; , \qquad u_2(t) = b \sin \Omega t \; . \tag{1.61}$$

Solution Neglecting the weak signal self-action, the exponent in (1.53) can be expanded into a series in terms of u_2. Restricting oneself to a linear term:

$$C(x, \omega) = \; c_0^2/(2\pi i \epsilon \omega x) \int_{-\infty}^{\infty} [e^{i(\epsilon/c_0^2)\omega x u_1(\xi)} - 1] \, e^{-i\omega \xi} \, d\xi +$$

$$\frac{1}{2\pi} \int_{-\infty}^{\infty} u_2(\xi) \, e^{i(\epsilon/c_0^2)\omega x u_1(\xi)} \, e^{-i\omega \xi} \, d\xi \; . \tag{1.62}$$

Here the first term describes the Fourier-transform of the pump wave while the second one is for the spectrum $C_{int}(x, \omega)$ resulting from the nonlinear interaction between the signal and the pump. Using relation (1.36) for Bessel functions and the filtering properties of the δ-function one obtains from (1.62)

$$C_{int}(x, \omega) = i\,b/2 \sum_{k=-\infty}^{\infty} \{J_k[\tfrac{\epsilon}{c_0^2}(k\omega_0 - \Omega)u_0 x]\delta(\omega - \Omega + k\omega_0) -$$

$$J_k[\tfrac{\epsilon}{c_0^2}(k\omega_0 + \Omega)u_0 x]\delta(\omega + \Omega + k\omega_0)\} \ (1.63)$$

Problem 1.23

Employing the result of the previous problem, consider the case of a low-frequency pump $\omega_0 \ll \Omega$ (this problem is a generalization of Problem 1.17). Describe the signal spectrum for various interaction stages and evaluate the signal spectrum width.

Solution For $\omega_0 \ll \Omega$ the nonlinear interaction leads to a high-frequency signal modulation and appearance of components at frequencies $\omega = \pm\Omega + k\omega_0$ $(k = 0, \pm1, \pm2, \dots)$ near the signal frequency. For the Fourier-transform (1.63) one obtains, respecting that $k\omega_0 \ll \Omega_L$,

$$C_{int}(x, \omega) = i\frac{b}{2} \sum_{k=-\infty}^{\infty} \{J_k(\frac{\Omega}{\omega_0}z)\delta(\omega - \Omega + k\omega_0) -$$

$$J_k(\frac{\Omega}{\omega_0}z)\delta(\omega + \Omega + k\omega_0)\} \ ,$$

$$\text{where} \quad z = \frac{\epsilon}{c_0^2}\omega_0 u_0 x = x/x_S \ . \tag{1.64}$$

It can be readily seen that (1.64) describes the spectrum of a signal with harmonic phase modulation

$$u_2(x, \tau) = b\sin[\Omega\tau + (\epsilon/c_0^2)\Omega x u_0 \sin \omega_0 \tau] \ . \tag{1.65}$$

Therefore, for $\Omega \gg \omega_0$ the interaction can be interpreted as a low-frequency phase modulation of a signal created by a powerful pump wave. As the waves propagate, the modulation depth increases. For $(\Omega/\omega_0)z \ll 1$ two harmonics $\omega = \Omega \pm \omega_0$ dominate in the spectrum. For $(\Omega/\omega_0)z \gg 1$, however, the spectrum broadens substantially. Using the Bessel function asymptotic behaviour for large arguments [22] one can estimate the effective harmonic number in the spectrum (1.64): $k_* \approx z(\Omega/\omega_0)$. The corresponding spectrum width is $\Delta\omega \sim z\Omega \gg \omega_0$.

Problem 1.24

Using the result of (1.63), consider the case of a high-frequency pump ($\Omega \ll \omega_0$). Interpret the nonlinear interaction process from a physical viewpoint.

Answer Nonlinear interaction leads in this case to appearance of two spectral components $\omega = k\omega_0 \pm \Omega$ near each one of the pump wave harmonics $\omega = k\omega_0$;

$$C_{int}(x,\omega) = i\frac{b}{2} \sum_{k=-\infty}^{\infty} J_k(kz)[\delta(\omega - \Omega + k\omega_0) - \delta(\omega + \Omega + k\omega_0)] \ . \ (1.66)$$

Chapter 2

PLANE NONLINEAR WAVES WITH DISCONTINUITIES

Problem 2.1

Determine the maximal distance, i.e. the range limits within which the solution (1.18) $u = \Phi(\tau + \epsilon ux/c_0^2)$ of the simple wave equation (1.13) holds.

Solution In Problem 1.8 the following derivative has been calculated

$$\frac{\partial u}{\partial \tau} = \frac{(\epsilon/c_0^2)u\,\Phi'}{1 - (\epsilon/c_0^2)x\,\Phi'} \ .$$ (2.1)

The maximal distance x_S is when the derivative (2.1) is infinite, which happens when

$$1 - \frac{\epsilon}{c_0^2}x\,\Phi'(\tau + \epsilon ux/c_0^2) = 0 \ .$$ (2.2)

Vanishing of denominator (2.2) in (2.1) corresponds to the fact that in a certain profile point at a distance x_S the derivative (2.1) turns to infinity, i.e. the tangent line to this point becomes vertical; in other words, this is the formation of a discontinuity (a shock). The sought profile point corresponds to the maximum value of function Φ', i.e. is obtained from the condition $\Phi'' = 0$. Therefore, the two conditions: $\Phi'' = 0$ and (2.2) will allow solving the posed problem. In practical situations it seems convenient to make use of the fact that a solution of the simple wave equation (1.13) can be written explicitly (1.25) with respect to $\tau(u, x)$

$$\tau = \Phi^{-1}(u) - \frac{\epsilon}{c_0^2}ux \ .$$ (2.3)

It has the following properties: a) at the distance $x = x_S$ a vertical tangent line to the curve $u(x_S, \tau)$ appears; b) a discontinuity is formed in the inflexion point leading to the following couple of equations

$$\frac{\partial \tau}{\partial u} = 0 \ , \qquad \frac{\partial^2 \tau}{\partial u^2} = 0 \ .$$ (2.4)

Problem 2.2

Solving the simple wave equation (1.13) by the method of characteristics presents an evident illustration of the single-valuedness condition formulated in the previous problem. Define which portion of the reference perturbation profile $u(x = 0, \tau) = \Phi(\tau)$ which will topple over first, and at what distance.

Solution A set of characteristic equations for the partial differential equation (1.13) has the form

$$\frac{\partial \tau}{\partial x} = -(\epsilon/c_0^2)u , \qquad \frac{\partial u}{\partial x} = 0 , \tag{2.5}$$

with $\tau(x = 0) = \tau_0$ and $u(x = 0, \tau_0) = \Phi(\tau_0)$. Here $\tau_0(u)$ is a point in the accompanying coordinate system from which a characteristic for the perturbation u comes out (Figure 2.1).

The solution of system (2.5)

$$\tau = \tau_0 - (\epsilon/c_0^2)\, \Phi(\tau_0)x \tag{2.6}$$

describes a family of straight lines in a plane (τ, x) with different slopes depending on $u = \Phi(\tau_0)$. It is worthwhile noting that (2.6) is the expression (2.3) written in other notations. The time interval between the neighbouring characteristics in accordance with (2.6) varies as

$$d\tau = d\tau_0[1 - (\epsilon/c_0^2)\Phi'(\tau_0)x] . \tag{2.7}$$

Therefore, the wave will topple over on the first intersection of characteristics (see Figure 2.1) and when $d\tau$ vanishes. This takes place at the distance

$$x_S = \frac{c_0^2}{\epsilon \cdot \max \Phi'(\tau_0)} . \tag{2.8}$$

The neighbourhood of the profile point in which a maximum of derivative Φ' is achieved will be the first to topple over.

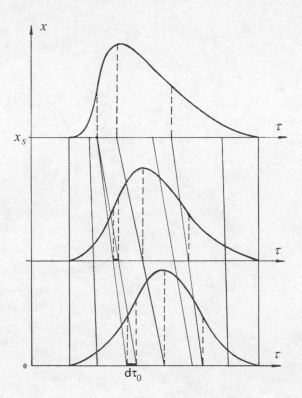

Figure 2.1: *Explanation of the turnover process of a Riemann wave.
The straight lines are the characteristics (2.6) and the intersection
of lines is evidence of the appearance of multi-valued dependence of
particle velocity on time.*

Problem 2.3

Find the distance at which a discontinuity is formed for a nonlinear
simple wave specified at the input in the form of the unipolar pulse
$u(x = 0, t) = u_0 e^{-t^2/t_0^2}$.

Solution *The first way*
Write the solution of the simple wave equation for the given unipolar
pulse as the explicit function $\tau = \tau(x, u)$:

$$\tau = -t_0 \sqrt{\ln(u_0/u)} - \epsilon u x/c_0^2 \ . \tag{2.9}$$

Here we use the minus before the root since a discontinuity is always formed in the leading front (in our case for $\tau < 0$).

It should satisfy the conditions (2.4):

$$\frac{\partial \tau}{\partial u} = \frac{t_0}{2u}[\ln \frac{u_0}{u}]^{-1/2} - \frac{\epsilon}{c_0^2}x = 0 , \qquad (2.10)$$

$$\frac{\partial^2 \tau}{\partial u^2} = -\frac{t_0}{2u^2}[\ln \frac{u_0}{u}]^{-1/2} + \frac{t_0}{4u^2}[\ln \frac{u_0}{u}]^{-3/2} = 0 . \qquad (2.11)$$

From (2.11) we find that a discontinuity is formed for $u = u_0/\sqrt{e}$. Substituting this value into (2.10) yields the distance $x_S = \sqrt{e/2}(c_0^2 t_0/\epsilon u_0)$.

The second way

Following the pattern described in Problem 2.2, calculate a derivative of the reference perturbation form

$$\Phi'(\tau_0) = -\frac{2\tau_0}{t_0^2}u_0 e^{-\tau_0^2/t_0^2} . \qquad (2.12)$$

A maximum of the function (2.12) is attained for $\tau_0 = t_0/\sqrt{2}$ and equals $\sqrt{2}u_0/\sqrt{e}\,t_0$. Formula (2.8) for this value immediately supplies the result x_S, which coincides with the expression obtained through the first way.

Problem 2.4

Find the coordinate x_S of the discontinuity formation in a harmonic reference wave $u(x = 0, t) = u_0 \sin \omega t$. Determine the profile points (τ) where discontinuities appear.

Answer A discontinuity is formed for $\omega \tau = 2\pi n$ at the distance $x_S = c_0^2/\epsilon \omega u_0$, $(n = 0, \pm 1, \pm 2, \ldots)$.

Problem 2.5

Find the coordinate of shock formation x_S in the step-wise disturbance $u(x = 0, t) = u_0 \tanh(t/t_0)$. Determine the profile point (τ) where the discontinuity appears.

Answer A discontinuity forms for $\tau = 0$ at the distance $x_S = c_0^2 t_0 / \epsilon u_0$.

Problem 2.6

A perturbation at the input is considered to be a superposition of harmonic oscillations with two incommensurable frequencies $u(x = 0, t) = u_1 \sin \omega_1 t + u_2 \sin \omega_2 t$. Determine the distance x_S at which the first discontinuity appears.

Answer The first discontinuity appears at the distance

$$x_S = \frac{c_0^2}{\epsilon(\omega_1 u_1 + \omega_2 u_2)} . \tag{2.13}$$

Problem 2.7

At which distance from a high-power ultrasound source does the shock form in water ? The intensity of the ultrasonic wave is $I = 10$ W/cm^2, the frequency is $f = 1$ MHz, the density is $\rho_0 = 1$ g/cm^3, the sound velocity is $c_0 = 1.5 \cdot 10^5$ cm/s, and the nonlinearity is $\epsilon = 4$.

Answer Using the formula given in the answer to Problem 2.4, the solution can be evaluated to be

$$x_S = \frac{c_0^2}{2\pi \cdot \epsilon \cdot f} \sqrt{\frac{c_0 \rho_0}{2I}} \approx 25 \text{ cm} . \tag{2.14}$$

Problem 2.8

Determine the intensity of a wave propagating in water for which a discontinuity forms at the distance 10 m. The frequency is $f = 200$ kHz.

Answer $I = 0.5 c_0^5 \rho_0 (2\pi \epsilon f x_S)^{-2} \approx 0.15$ W/cm^2.

Problem 2.9

Evaluate the vibration velocity amplitude, the particle displacement, the particle acceleration, and the magnitude of the acoustic Mach number for Problems 2.7 and 2.8.

Answer For Problem 2.7: velocity $u_0 = (2I/c_0\rho_0)^{1/2} \approx 36$ cm/s, displacement $\xi_0 = u_0/\omega \approx 6 \cdot 10^{-6}$ cm, acceleration $a_0 = \omega u_0 \approx 2 \cdot 10^8$ cm/s^2, and Mach number $M = u_0/c_0 \approx 2.4 \cdot 10^{-4}$.

For Problem 2.8: $u_0 \approx 4.5$ cm/s, $\xi_0 \approx 4 \cdot 10^{-6}$ cm, acceleration $a_0 \approx 5 \cdot 10^6$ cm/s^2, and Mach number $M \approx 3 \cdot 10^{-5}$.

One can see, that the displacements of particles are very small even in high-power ultrasonic fields; they are on the order of molecular scales. On the other hand, huge accelerations exist, up to 10^6 g, where g is acceleration of gravity on earth (g≈ 9.8 m/s^2). Mach numbers are small, and this fact was already used at the simplification of the nonlinear equations in Problems 1.2 and 1.5.

Problem 2.10

A plane monochromatic wave propagates in air. Write the formula for the shock formation distance through the sound pressure level N(dB) and frequency f. Evaluate the Mach number and shock formation distance for $N = 140$ dB (this level corresponds to the noise of a heavy jet aircraft) and $f = 3300$ Hz. The adiabatic power index for air is $\gamma = 1.4$.

Solution It is common for atmospheric acoustics to characterize the sound intensity by the level of root-mean-square pressure N(dB) relative to the reference pressure $p_* = 2 \cdot 10^{-5}$ Pa. The peak pressure is $p' = \sqrt{2}p_* 10^{N/20}$. The shock formation distance for a plane monochromatic wave is determined by the formula (1.16), where $M = u_0/c_0$, and u_0 is the peak magnitude of vibrational velocity. With account for the definition $c_0^2 = \gamma p_0/\rho_0$, where ρ_0 is density of air and p_0 is atmospheric pressure ($p_0 \approx 10^5$ Pa), we write the Mach number in the form $M = p'/(c_0^2\rho_0) = p'/(\gamma p_0)$. Consequently, the Mach number and the shock formation distance are evaluated as

$$M = \sqrt{2} \cdot \frac{p_*}{\gamma p_0} 10^{N/20} \approx 2 \cdot 10^{-3} \ ,$$

$$x_S = \frac{c_0}{f} \cdot \frac{\gamma}{\gamma+1} \cdot \frac{1}{\pi\sqrt{2}} \cdot \frac{p_0}{p_*} 10^{-N/20} \approx 6 \text{ m} \ . \qquad (2.15)$$

Problem 2.11

Make a simple geometrical construction based on the conservation of momentum transferred by a simple wave which removes the non-single-valuedness of the overhang profile at distances $x > x_S$ (Figure 2.2).

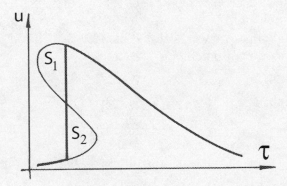

Figure 2.2: *Construction of a shock from a multi-valued profile of a Riemann wave in accordance with the rule of equal areas.*

Solution Make sure that the momentum in a limited simple wave ($u \to 0$ for $\tau \to \pm\infty$) is independent of x for $x < x_S$

$$\int_{-\infty}^{+\infty} \rho_0 u \, d\tau = \rho_0 \int_{-\infty}^{+\infty} \Phi(\tau + \frac{\epsilon}{c_0^2} ux) \, d\tau =$$

$$\rho_0 \int_{-\infty}^{+\infty} \Phi(\xi) \, d(\xi - \frac{\epsilon}{c_0^2} x\Phi) = \rho_0 \int_{-\infty}^{+\infty} \Phi(\xi) \, d\xi \ . \qquad (2.16)$$

The geometrical meaning of the conservation law is the keeping of a constant area between the wave profile $\Phi(x, \tau)$ and the τ axis. After the overhang formation ($x > x_S$), this area has to be retained inasmuch as the medium region taken up by the wave motion remains closed (not affected by external forces). Consequently, a discontinuity in a non-single-valued wave profile has to be introduced such that the cut-off areas S_1 and S_2 (see Figure 2.2) are equal. In fact, the area S_1 is added to the profile whereas the area S_2 is subtracted from it. Provided that $S_1 = S_2$, the area under the curve is equal to the reference value $\int_{-\infty}^{+\infty} \Phi(\xi)\, d\xi$.

Problem 2.12

Demonstrate that a compressive shock wave, a jump between the two constant values u_1 and u_2 with $u_2 > u_1$, is stable, i.e. is not supposed to change its form during propagation.

Solution Let for simplicity $u_1 = 0$, $u_2 > 0$. At a reference point $x = 0$ a wave has a discontinuity located at $\tau = 0$ within the accompanying coordinate system. For distances $x > 0$ the distorted profile is constructed graphically by the method outlined in Problem 1.10. It is evident that the profile becomes non-single-valued for arbitrary small x (dashed line in Figure 2.3a). This non-single-valuedness is removed in accordance with the rule of "equality of areas" (see Problem 2.11). As a result one obtains a jump of the same shape and magnitude but with a front slightly displaced forward. This means that the compression wave is stable. The front displacement within the accompanying coordinate system $\tau = t - x/c_0$ testifies that the positive (with respect to the unperturbed level $u_1 = 0$) jump u_2 travels with supersonic speed $c = c_0 + \epsilon u_2/2$ - the faster the higher the shock jump. It is worthwhile noting that the rarefaction shock wave ($u_2 < u_1$) is unstable, i.e. its front thickness increases as the wave propagates (Figure 2.3b). One only needs to make use of the graphic technique described in Problem 1.10 to confirm this. The rule of "equality of areas" is not required in this case.

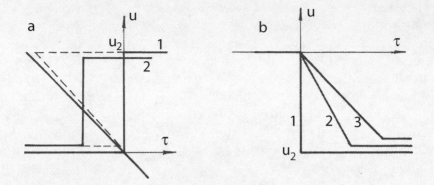

Figure 2.3: *Evolution of a compression shock wave (a) and a rarefaction wave (b). The graphical analysis shows the compression shock stability, but the rarefaction shock spreads out under the influence of nonlinearity*

Problem 2.13

Using the area equality rule, define the location and amplitude of a discontinuity $u_S(x)$ of the sinusoidal reference perturbation $u(x = 0, t) = \Phi(t) = u_0 \sin \omega t$. Determine the distance x_* at which the value of $u_S(x)$ is maximal and formulate an asymptotic law of its variation for large x.

Answer Within the moving coordinate system $\tau = t - x/c_0$, the discontinuities (one for each period) take the fixed positions at $\omega \tau = 2\pi n$ $(n = 0, \pm 1, \pm 2, \ldots)$ and their amplitude is defined as a nonzero root of the equation $\arcsin(u_S/u_0) = z(u_S/u_0)$ where $z = x/x_S = (\epsilon/c_0^2)\omega u_0 x$. A maximum value of $u_S(x_*) = u_0$ is attained for $z = z_* = \pi/2$. An asymptotic decreasing law $u_S/u_0 = \pi/(1 + z)$ is observed for $z > 2$. It is of interest that for $z \gg 1$, $u_S \approx c_0^2/\epsilon\omega x$, which does not depend on the input signal amplitude. The solution to the problem is set forth in [4] and [6].

Problem 2.14

Using the results of the previous problem, find the shape of a sinusoidal input at the distances $z = x/x_S > 2$. Calculate the spectral composition and the energy density average over the period: $E = \rho_0 \overline{u^2} = \rho_0 T^{-1} \int_0^T u^2(x, \tau) \, d\tau$.

Answer The wave acquires the sawtooth profile

$$\frac{u}{u_0} = \frac{1}{1+z}(-\omega\tau + \pi \operatorname{sign}(\tau)) , \qquad -\pi \le \omega\tau \le \pi . \tag{2.17}$$

Its spectrum is

$$\frac{u}{u_0} = \sum_{n=1}^{\infty} \frac{2}{n(1+z)} \sin n\omega\tau . \tag{2.18}$$

Due to discontinuity formation and their nonlinear damping (which grows higher for larger u_0), the harmonic amplitudes decrease following the power law $A_n \sim n^{-1}$. The energy density decreases as $E = \pi^2 \rho_0 u_0^2 / 3(1+z)^2$ and for $z \gg 1$ it does not depend on the reference perturbation amplitude u_0.

Problem 2.15

Making use of the plottings described in Problems 1.10 and 2.11, follow the evolution of the rectangular-at-input pulse $\Phi(\tau) = A$ for $-T < \tau < 0$, and $\Phi(\tau) = 0$ outside this interval. Find an asymptotic form of the pulse as $x \to \infty$.

Answer The initial shape of the pulse and its shape for three characteristic distances are depicted in Figure 2.4.

For $x(\epsilon/c_0^2)A/T \gg 1$ the pulse acquires a universal triangular shape with a slope which is independent of A and T:

$$u = -\frac{c_0^2 \tau}{\epsilon x} , \quad (-\tau_S < \tau < 0) \qquad u = 0 , \quad (\tau < \tau_S , \ \tau > 0) . \tag{2.19}$$

Here $\tau_S(x) = [2AT(\epsilon/c_0^2)x]^{1/2}$ is the current pulse duration. It can be proven that for any x the pulse area is equal to AT, which is in line with the conservation of momentum.

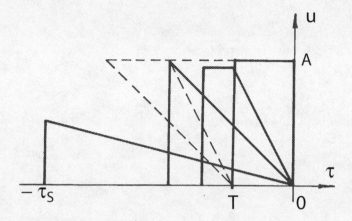

Figure 2.4: **Problem 2.15** *Evolution of the shape of an initially rectangular pulse. Because of the motion of the leading shock front the pulse becomes triangular. Thereafter, the peak value decreases and an increase in its duration takes place.*

Problem 2.16

Perform a graphical analysis of the nonlinear evolution of a bipolar sound pulse consisting of two symmetric triangular pulses (see Problem 1.11) with duration $2T_0$ and area S for the following cases: a) a rarefaction phase followed by a compression phase; b) a compression phase followed by a rarefaction phase.

Answer It is evident from Figure 2.5, that, for case a) the pulse is transformed into a so-called S-wave of constant duration $2T_0$; and, for case b) the pulse is transformed into an N-wave duration of which $2T(x)$ grows with x. Increase of the curve number in Figure 2.5 corresponds to increase of distance.

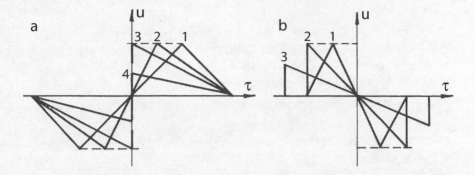

Figure 2.5: **Problem 2.16** *The evolution of the shape of a bipolar pulse is different in the compression and rarefaction areas. A pulse can transform into an S-wave (a) or to an N-wave. Higher number of curve corresponds to increase in the distance traveled by the wave.*

Problem 2.17

Find the asymptotic behaviour of the Fourier-transforms for $(\epsilon/c_0^2)Sx/T_0^2 \gg 1$ in the previous problem's conditions by employing the linear profile evolution results in Problem 1.9. Discuss characteristic properties of the spectra structure in the high and low frequency ranges.

Answer a) The spectrum of the S-wave:

$$|C(x,\omega)| = \frac{2T_0^2}{\omega T_0}(1 - \frac{\sin\omega T_0}{\omega T_0})\frac{c_0^2}{\epsilon x} . \qquad (2.20)$$

b) The N-wave spectrum is self-similar:

$$|C(x,\omega)| = \frac{2T^2}{\omega T}|\cos\omega T - \frac{\sin\omega T}{\omega T}|\frac{c_0^2}{\epsilon x} , \qquad T = [2S\frac{\epsilon}{c_0^2}x]^{1/2} . \qquad (2.21)$$

Spectra (2.20) and (2.21) are displayed in Figure 2.6.

Within the high-frequency range the spectra drop according to the power law which is associated with the presence of discontinuities.

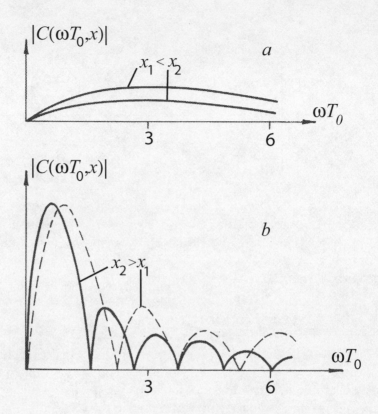

Figure 2.6: *Asymptotic behavior of the spectra of bipolar pulses at large distances in a nonlinear medium of initial: S-wave (a) and N-wave (b).*

For the S-pulse all the spectrum components diminish as $\sim 1/x$. For the N-pulse the spectrum maximum is observed to gradually move in the direction of low frequencies; at high frequencies the spectral components decrease as $1/\sqrt{x}$, whereas for low frequencies it looks like $|C(x,\omega)| \sim \omega\sqrt{x}$. The spectral density growth at low frequencies is related to the parametric energy pumping under nonlinear interaction of high frequency harmonics.

Problem 2.18

Formulate a set of equations describing the evolution of the profile of a simple wave containing a discontinuity.

Solution Obtain a differential equation describing the shock front motion within the accompanying coordinate system. Consider a discontinuity at a distance x which has the coordinate $\tau_S(x)$ (Figure 2.7).

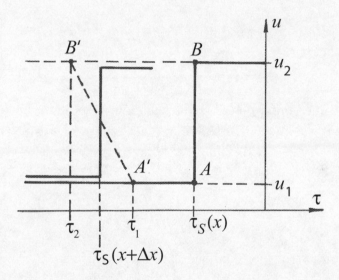

Figure 2.7: *In the derivation of equation (2.23) the rule of equal areas describing the shock motion within the Riemann wave profile is used.*

The vibration velocity immediately before the front (point A) is u_1, and immediately after the front (point B) it is u_2. When the distance increases by Δx, the point A moves to A' which has the coordinate $\tau_1 = \tau_S(x) - (\epsilon/c_0^2)u_1\Delta x$ and the point B moves to B' with the coordinate $\tau_2 = \tau_S(x) - (\epsilon/c_0^2)u_2\Delta x$. The rule of equal area (see Problem 2.11) implies that the new coordinate of the discontinuity

$$\tau_S(x + \Delta x) = \frac{1}{2}(\tau_1 + \tau_2) = \tau_S(x) - \frac{\epsilon}{2c_0^2}(u_1 + u_2)\Delta x \ . \qquad (2.22)$$

Passing to the limit in (2.22) as $\Delta x \to 0$, we come to the equation

$$\frac{d\tau_S}{dx} = -\frac{\epsilon}{2c_0^2}(u_1 + u_2) \ . \tag{2.23}$$

Therefore, the front travel speed in the accompanying coordinate system depends only on the perturbation values u_1 and u_2 at the discontinuity which, generally speaking, depend on the distance x. Inasmuch as u_1 and u_2 not only belong to the discontinuity but to the simple wave profile as well, the solution in (2.3) will be valid for them, i.e.

$$\tau_S(x) = \Phi_1^{-1}(u_1) - \frac{\epsilon}{c_0^2}u_1 x \ , \tag{2.24}$$

$$\tau_S(x) = \Phi_2^{-1}(u_2) - \frac{\epsilon}{c_0^2}u_2 x \ . \tag{2.25}$$

Here function Φ_1 describes the simple wave profile before the discontinuity, and Φ_2 for after the discontinuity. $\Phi_{1,2}^{-1}$ are the inverse functions of $\Phi_{1,2}$. The three equations (2.23) through (2.25) for the three unknowns $\tau_S(x)$, $u_1(x)$, $u_2(x)$ furnish a complete set for solving the problem.

Problem 2.19

Making use of the equations (2.23) through (2.25) in the previous problem, find the change with distance of the jump magnitude and the duration of a triangular pulse with a shock wave at the leading edge. For $x = 0$ the pulse is prescribed as follows: $u/u_0 = 1 - \tau/T_0$ $(0 < \tau < T_0)$, $u_0 = 0$ for all remaining τ.

Answer The jump value decrease is accompanied by the duration rise

$$\frac{u_2(x)}{u_0} = (1 + \frac{\epsilon u_0}{c_0^2 T_0}x)^{-1/2} \qquad T(x) = T_0(1 + \frac{\epsilon u_0}{c_0^2 T_0}x)^{1/2} \ . \tag{2.26}$$

Since momentum remains the same, the pulse area $u_2(x)T(x) = u_0 T_0 = $ constant.

Problem 2.20

Demonstrate that two following weak shock waves collide according to the law of absolutely inelastic impact of two particles. The jump magnitude $(u_2 - u_1)$ is used as an analogue of the particle mass m and the velocity $d\tau_S/dx = -(\epsilon/2c_0^2)(u_1 + u_2)$ of the front within the accompanying coordinate system is an analogue of the particle velocity v.

Solution Using the methods of graphical analysis one can demonstrate that two shock waves, i.e. the perturbation discontinuities $(u_2 - u_1)$ and $(u_3 - u_2)$ (Figure 2.8) merge to form a single wave with difference $(u_3 - u_1)$.

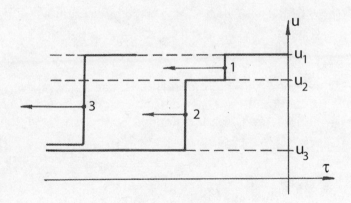

Figure 2.8: *The junction of two shocks is analogous to the inelastic collision of two balls.*

The trivial relationship

$$(u_3 - u_1) = (u_2 - u_1) + (u_3 - u_2) \tag{2.27}$$

is an analogue of the particle mass conservation law: $m' = m_1 + m_2$. As an analogue of the law of conservation of momentum $m'v' = m_1v_1+$

$m_2 v_2$, the following relationship can be offered

$$(u_2 - u_1)\frac{\epsilon}{2c_0^2}(u_1 + u_2) + (u_3 - u_2)\frac{\epsilon}{2c_0^2}(u_2 + u_3) =$$

$$= (u_3 - u_1)\frac{\epsilon}{2c_0^2}(u_1 + u_3) , \qquad (2.28)$$

which is seen to be an identity.

Problem 2.21

A weak shock wave, described by the function $u = \Phi(\tau + \epsilon u x/c_0^2)$, propagates in a nonperturbed medium. Find the distance dependence of the jump value at the shock wave front.

Solution Make use of the pair of equations obtained in Problem 2.18:

$$\tau_S = \Phi^{-1}(u_2) - \frac{\epsilon}{c_0^2}u_2 x , \qquad \frac{d\tau_S}{dx} = -\frac{\epsilon}{2c_0^2}u_2 . \qquad (2.29)$$

Here $\tau_S(x)$ describes the discontinuity position in the accompanying coordinate system, and $u_2(x)$ is the jump value. Eliminating $\tau_S(x)$ from the equations (2.29) and assuming $x = x(u_2)$ yield the linear equation

$$\frac{1}{2}u_2\frac{dx}{du_2} + x = \frac{c_0^2}{\epsilon}\frac{d}{du}\Phi^{-1}(u_2) . \qquad (2.30)$$

Solving (2.30) the general expression is obtained as [12]:

$$\frac{\epsilon}{2c_0^2}x(u_2) = \frac{1}{u_2^2}\int^{u_2} u \, d\Phi^{-1}(u) . \qquad (2.31)$$

The integration constant can be chosen, for example, by the initial coordinate x_S of the discontinuity formation (see Problems 2.1 - 2.6): $x(u_2^*) = x_S$, where $u_2^* = u_2(x_S)$ is the initial jump value (zero, as a rule, unless the leading edge of the reference simple wave is a straight line segment).

Problem 2.22

Using formula (2.31) of the previous problem, find the amplitude for the discontinuity $u_2(x)$ when a solitary pulse equal to $u = u_0 \sin \omega\tau$ for $0 \leq \omega\tau \leq \pi$, and $u = 0$ for all other $\omega\tau$, propagates.

Solution In this problem $\Phi^{-1} = \omega^{-1} \arcsin(u/u_0)$ and formula (2.31) acquires the form

$$\frac{\epsilon\omega}{2c_0^2}x = \frac{1}{u_2^2} \int^{u_2} \frac{u}{u_0}\left(1 - \frac{u^2}{u_0^2}\right)^{-1/2} du . \tag{2.32}$$

Calculating the integral yields

$$U^{-2}(\sqrt{1 - V^2} + C) = -x/2 , \tag{2.33}$$

where $V = u_2(x)/u_0$, $z = (\epsilon/c_0^2)\omega u_0 x$, and C is the integration constant. A calculation as in Problem 2.4 follows, with $z_S = 1$ and a discontinuity formation starting from a zero-amplitude jump $V(z_S = 1) = 0$. Because of this the constant $C = -1$. Therefore, the shock wave amplitude variation in space follows the law $u_2(x) = 0$ for $x < x_S$ and

$$u_2(x)/u_0 = 2\sqrt{z - 1}/z \tag{2.34}$$

for $z > z_S = 1$ (or $x > x_S$). The discontinuity amplitude is observed to increase for $1 < z < 2$, attaining a maximum value $u_2 = u_0$ for $z = 2$ followed by a drop $\sim 1/\sqrt{z}$ (in the region of $z > 2$).

Problem 2.23

Find the change of duration of a solitary pulse, i.e. a sinusoidal half-period (see the previous problem).

Solution Substitution of the solution (2.34) to the equations (2.29) leads to following formulas:

$$\frac{d}{dz}(\omega\tau_S) = -\frac{V}{2} = -\frac{\sqrt{z - 1}}{z} ,$$
$$\omega\tau_S = C + 2\arctan\sqrt{z - 1} - 2\sqrt{z - 1} . \tag{2.35}$$

Because the shock starts to form at $z = 1$ in the profile point $\tau_S = 0$, the constant $C = 0$. Consequently, the pulse duration

$$T = \frac{\pi}{\omega} , \qquad (z < 1) ;$$

$$T = \frac{\pi}{\omega} + \frac{2}{\omega}(\sqrt{z-1} - \arctan\sqrt{z-1}) , \quad (z > 1) , \qquad (2.36)$$

is constant before the shock formation and increases monotonically after shock forms because it moves with variable supersonic speed.

Problem 2.24

This problem, and those which follow (2.24-2.28), are based on the results obtained by D.G. Crighton[2].

They are concerned with simple waves in a cubically nonlinear medium whose propagation is governed by the equation (1.13)

$$\frac{\partial u}{\partial x} = \gamma u^2 \frac{\partial u}{\partial \tau} . \qquad (2.37)$$

Find a solution to this equation and perform a graphic analysis with respect to a sinusoidal signal.

Solution Equation (2.37) can be solved by the method of characteristics. Or, by means of a direct substitution one can verify that it is satisfied by the implicit expression

$$u(x, \tau) = \Phi(\tau + \gamma u^2 x) . \qquad (2.38)$$

In the case of a sinusoidal signal we write

$$\omega\tau = \arcsin(u/u_0) - \gamma\omega u_0^2 x(u/u_0)^2 . \qquad (2.39)$$

In order to graphically analyze the nonlinear distortion of the wave profile, make use of the expression (2.39) and proceed in analogy with

[2]See original paper: I.P. Lee-Bapty, and D.G. Crighton. Phil.Trans.Roy.Soc.London, **A 323**, 173-209, 1987; and the generalization for diffracting beams: O.V. Rudenko, and O.A. Sapozhnikov, J.Exp.Theor.Phys.(JETP) **79**(2), 220-228, 1994.

Problem 1.10. As distinct from the quadratically nonlinear medium for which the distorted profile is achieved through graphic summation (see Figure 1.1) of the reference profile and the straight line, here - in line with (2.39) - the straight line has to be replaced with a parabola. The appropriate construction is carried out in Figure 2.9. A break in a non-single-valued profile was effected following the rule of the area equality; the cut-off portions are cross-hatched. It is evident that the wave profile acquires a saw-tooth shape. Unlike for a quadratically-nonlinear medium, however, where the "saw" is composed of triangular teeth in the cubic medium the "teeth of the saw" reminds of a trapezoid.

Problem 2.25

Find the coordinate of the discontinuity formation in a sinusoidal reference wave propagating in a cubically-nonlinear medium. Identify the profile points where the discontinuities appear.

Solution Let us employ formula (2.39) of the previous problem. Denote for convenience $V = u/u_0$, $\theta = \omega\tau$, and $z = \gamma\omega u_0^2 x$:

$$\theta = \arcsin V - zV^2 \ . \tag{2.40}$$

Proceeding as in Problem 2.3 (the first method) and considering for simplicity the first half-period $(0 < \theta < \pi)$ we arrive at

$$\frac{\partial\theta}{\partial V} = (1 - V^2)^{-1/2} - 2zV = 0 \ , \qquad \frac{\partial^2\theta}{\partial V^2} = V(1 - V^2)^{-3/2} - 2z = 0 \ . \tag{2.41}$$

These two equations imply that a discontinuity starts to form in the profile point where $V = 1/\sqrt{2}$ at a distance of $z = 1$.

Problem 2.26

Formulate a differential equation of the type (2.23) describing the discontinuity propagation in a cubically nonlinear medium.

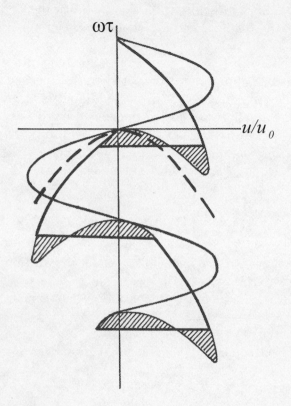

Figure 2.9: *Graphical analysis of the distortion of a Riemann (simple) wave in a cubically nonlinear medium.*

Solution Consider a non-single-valued portion of the profile depicted in Figure 2.10.

According to the law of conservation of momentum the shaded areas S_1 and S_2 have to be equal, that is

$$\frac{d}{dx} \int_{u_1}^{u_2} [\tau(u) - \tau_S(x)] \, du = 0 \ , \tag{2.42}$$

where u_1 is the value of the particle velocity before the front, and u_2 is the value immediately after the front. Substituting the solution of

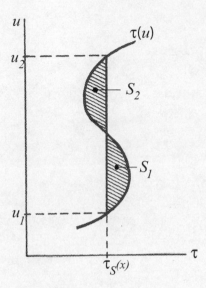

Figure 2.10: *The position of shock in a cubically nonlinear medium used in the derivation of equation (2.45).*

the cubic simple wave equation (2.37) in the form

$$\tau = \Phi^{-1}(u) - \gamma u^2 x \tag{2.43}$$

into (2.42) and calculating the integral one obtains

$$(-\frac{\gamma}{3}u^3 - \frac{d\tau}{dx}u)\Big|_{u_1}^{u_2} = 0 . \tag{2.44}$$

Hence the desired result follows:

$$\frac{d\tau_S}{dx} = -\frac{\gamma}{3}(u_1^2 + u_2^2 + u_1 u_2) . \tag{2.45}$$

The front propagation velocity within the accompanying coordinate system governed by equation (2.45) depends only on the values of u_1, u_2 of the discontinuity (which, generally, can depend on x). However, in contrast to the equation for a quadratically nonlinear medium (2.23), equation (2.45) describes a more complex motion. Therefore, inasmuch as $u_1^2 + u_2^2 + u_1 u_2 > 0$ for any value of u_1, u_2 the motion of any discontinuity across a cubically nonlinear medium is always with supersonic speed (naturally, as becoming more rigid in nonlinear media for which $\gamma > 0$). Among all discontinuities under fixed u_2 the slowest is that for which $u_1 = -u_2/2$; its motion is described by the equation $d\tau_S/dx = -\gamma(u_2/2)^2$.

Problem 2.27

Using graphic analysis of the profile (see Problem 2.24) and equation (2.45), demonstrate that in a cubically nonlinear medium a "slow" jump for which $u_1 = -u_2/2$ and $u_2 > 0$ (see the previous problem) is stable. Show also that a jump with a perturbation value before the shock wave front equal to $u_1 < -u_2/2$ is unstable, i.e. its shape gets distorted on propagation.

Solution Substituting the values of $u_1 = -u_2/2$ and $u_2 > 0$ into equation (2.45) yields $d\tau_S/dx = -\gamma(u_2/2)^2$. This means that the front coordinate $\tau_S(x)$ is

$$\tau_S(x) = -\gamma(u/2)^2 x \tag{2.46}$$

at a distance x (provided that $\tau_S(0) = 0$), coinciding with the position of a point A on the simple wave profile which corresponds to the perturbation $u_1 = -u_2/2$ before the front. As is shown in Figure 2.11 the advanced front passes exactly over the point A belonging to parabola $\tau = -\gamma u^2 x$.

Because of this the wave at a distance still has the form of a jump between the values of $u_1 = -u_2/2$ and u_2. Thus, this wave is stationary, i.e. its form does not change on propagation.

As is evident from the similar construction in Figure 2.12, the reference jump between the values of $u_1 < -u_2/2$ and $u_2 > 0$ is unstable. An AB portion of the simple wave profile emerges before the front and the form of the whole wave at a distance x differs from the reference step.

It has to be noted that a rarefaction jump with the following parameters at discontinuity: $u_1 > 0$, $u_2 = -u_1/2$ turns out to be a stable

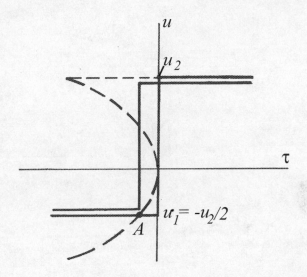

Figure 2.11: *A stable "slow" compression shock in a cubically nonlinear medium.*

"slow" wave. This fact results from the equation (2.45) invariance with respect to the transformation $u_1 = -u_2$, $u_2 = -u_1$ or, more generally, from the invariance of (2.37) with respect to the substitution $u \to -u$.

Problem 2.28

Demonstrate that equation (2.37) for the simple waves in a cubically nonlinear medium has a solution of the type

$$u(x, \tau) = f(x)\, \Phi(\tau - \tau_S(x)) \ . \tag{2.47}$$

Define unknown functions which describe: $f(x)$ - the variation of the characteristic wave amplitude; and, $\tau_S(x)$ - the wave motion as a whole in the accompanying coordinate system. Find also the stationary profile shape, i.e. the function $\Phi(\tau)$.

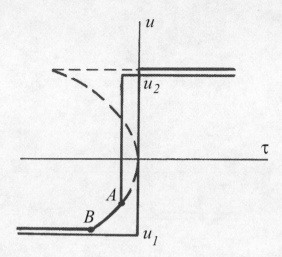

Figure 2.12: *An unstable compression shock in a cubically nonlinear medium*

Solution Substituting (2.47) into the simple wave equation gives the relationship

$$f' \, \Phi - f \, \tau_S' \frac{d\Phi}{d\tau} = \gamma f^3 \Phi^2 \frac{d\Phi}{d\tau} \ . \tag{2.48}$$

In (2.48) the variables have to be separated. This is possible provided the x-dependent functions are equal to

$$f = u_0 (1 + x/x_0)^{-1/2} \ , \qquad \tau_S = -\tau_0 \ln(1 + x/x_0) \ . \tag{2.49}$$

Here x_0 and τ_0 are arbitrary positive constants. In deducing the functions $f(x)$ and $\tau_S(x)$ the following boundary conditions were employed for definiteness: $f(0) = u_0$ - the reference amplitude is equal to u_0; and, $\tau_S(0) = 0$ - the reference front position is the origin of the accompanying coordinate system. Substituting the obtained solutions of (2.49) into (2.48) yields an ordinary differential equation for the function Φ

$$\left(\frac{2\tau_0}{\Phi} - 2\gamma u_0^2 x_0 \Phi\right) \frac{d\Phi}{d\tau} = 1 \ . \tag{2.50}$$

The solution of (2.50) has the form

$$\tau = C + 2\tau_0 \ln |\Phi| - \gamma u_0^2 x_0 \Phi^2 \ . \tag{2.51}$$

As the formula (2.51) defines a non-single-valued and unbounded function $\Phi(\tau)$, solution (2.51) can not correspond to a wave possessing physical meaning. However, an arbitrary integration constant C in the solution (2.51) permits to displace the function $\Phi(\tau - C)$ along the τ axis and obtain a periodic sequence of branches. Connecting these branches by discontinuities, D.G. Crighton managed to construct a periodic trapezoidal "saw" (see Figure 2.13) representing an asymptotic form (under $z = \gamma \omega u_0^2 x \gg 1$) of the reference sinusoidal signal (see Figure 2.9).

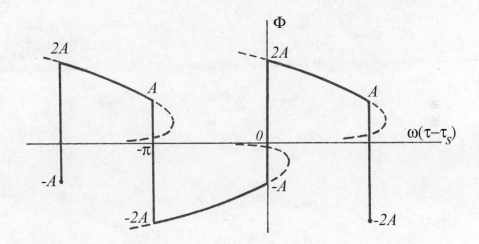

Figure 2.13: *The asymptotic (at large distances) profile of a periodic signal having sinusoidal shape at the input of a cubically nonlinear medium. Each saw-tooth is similar to a trapezium.*

It is worthwhile noting that the fronts depicted in Figure 2.13 appear as an alternating sequence of two discontinuities with parameters $u_1 = -A$, $u_2 = 2A$ and $u_1 = -2A$, $u_2 = A$. Both the discontinuities, as is shown in Problem 2.27, are "slow" stable shock waves propagating with the same velocity.

Chapter 3

NONLINEAR WAVES IN DISSIPATIVE MEDIA AND BURGERS' EQUATION

Problem 3.1

Making use of the method of slowly varying profile (see Problem 1.5), simplify the linear equation

$$\frac{\partial^2 u}{\partial t^2} - c_0^2 \frac{\partial^2 u}{\partial x^2} = \frac{b}{\rho_0} \cdot \frac{\partial^3 u}{\partial t \, \partial x^2} \, , \tag{3.1}$$

describing propagation of sound in a viscous heat-conducting medium (see [4],[6]). Here $\beta = 4\eta/3 + \xi + \kappa(c_v^{-1} - c_p^{-1})$ is a dissipation factor where ξ, η are bulk and shear viscosity, respectively, and κ is thermal conductivity. Obtain a solution to the resulting equation for sinusoidal and unipolar pulsed input signals.

Solution It is assumed that dissipative effects lead to slow distortion of the profile which allows one to pass over to the accompanying coordinates $\tau = t - x/c_0$, $x_1 = \mu x$. The terms μ^2, μ^3, ... are disregarded while the terms of the order of μ^0 cancel each other. As a result, the terms of the same infinitesimal order μ^1 remain, which make up the parabolic type equation

$$\frac{\partial u}{\partial x} = \delta \frac{\partial^2 u}{\partial \tau^2} \, , \qquad \delta = \frac{b}{2c_0^3 \rho_0} \, . \tag{3.2}$$

The general solution of (3.2) conforming to the arbitrary shape initial perturbation $u(x = 0, t) = u_0(t)$ is expressed with the aid of the Green function

$$u(x, \tau) = \int_{-\infty}^{+\infty} u_0(\tau') G(x, \tau - \tau') \, d\tau' \, , \qquad G = \frac{\exp(-\tau^2/4\delta x)}{\sqrt{4\pi\delta x}} \, . \tag{3.3}$$

For the harmonic perturbation $u_0 = a \sin(\omega t)$ one obtains

$$u(x, \tau) = a \exp(-\delta \omega^2 x) \sin(\omega t) \tag{3.4}$$

which is an exponentially damped wave. The quantity x_d is the inverse of the damping factor $(x_d \sim 1/(\delta \omega^2))$ and is called characteristic damping length. The condition $x_d \gg \lambda$ implies that the amplitude of the wave (3.4) drops insignificantly over distances on the order of the wave length λ. The relation

$$\lambda/x_d = \frac{\pi b \omega}{c_0^2 \rho_0} \sim \mu \ll 1 \; , \tag{3.5}$$

is a small parameter of the problem; it is of the order of the ratio of the right-hand side in (3.1) to any one of the terms on the left-hand side. The presence of the small parameter μ justifies the transition from (3.1) to (3.2).

For a unipolar pulse having characteristic duration t_0 at the input, the width of the function $G(x, \tau)$ is observed to greatly exceed t_0 at the distances $4\delta x/t_0^2 \gg 1$ and formula (3.3) is simplified to

$$u(x \gg t_0^2/4\delta, \tau) = G(x, \tau) \int_{-\infty}^{\infty} u_0(\tau') \, d\tau' \; . \tag{3.6}$$

Therefore, for large distances the pulse takes on the asymptotic form of the Gaussian curve.

Problem 3.2

Deduce an evolution Burgers equation describing the slow processes of the wave profile distortion due to nonlinear and dissipative properties of the medium.

Solution Using the method of slowly varying profile, the simple wave equations (1.13) and the parabolic equation (3.2) were derived earlier

$$\frac{\partial u}{\partial x} = \beta u \frac{\partial u}{\partial \tau} \; , \qquad \frac{\partial u}{\partial x} = \delta \frac{\partial^2 u}{\partial \tau^2} \; , \qquad (\beta = \epsilon/c_0^2, \; \delta = b/(2c_0^3 \rho_0)) \; , \tag{3.7}$$

describing the profile evolution due to nonlinear and dissipative effects separately. Inasmuch as these effects are weak they are described by independent terms in the reference equations. As a consequence, non-linear and dissipative terms will enter the simplified equation additively in the form of individual summands. Thus, a generalization of equations (3.7) results in

$$\frac{\partial u}{\partial x} = \beta u \frac{\partial u}{\partial \tau} + \delta \frac{\partial^2 u}{\partial \tau^2} , \qquad (3.8)$$

which is called Burgers equation. If we pass in (3.8) to the dimension-less variables

$$V = u/u_0 , \qquad \theta = \omega\tau , \qquad z = \beta\omega u_0 x = x/x_S \qquad (3.9)$$

- where u_0 is the characteristic perturbation value (for example, the harmonic wave amplitude or peak perturbation in the pulse) and ω is a characteristic frequency of a periodic signal (or the inverse pulse duration) - the equation acquires the form

$$\frac{\partial V}{\partial z} = V \frac{\partial U}{\partial \theta} + \Gamma \frac{\partial^2 V}{\partial \theta^2} . \qquad (3.10)$$

Here the number

$$\Gamma = \frac{b\omega}{2\epsilon c_0 \rho_0 u_0} = \frac{1}{2\epsilon \mathrm{Re}} = \frac{\omega}{u_0} \cdot \frac{\delta}{\beta} , \qquad (3.11)$$

is a single dimensionless set of parameters entering equation (3.8), thereby completely defining the evolution process.

Sometimes, the acoustic Reynolds number $\mathrm{Re} = (2\epsilon\Gamma)^{-1}$ is used instead of Γ. One can write Γ as a relation between characteristic nonlinear and dissipative lengths;

$$\Gamma = \frac{x_S}{x_d} = \frac{c_0^2}{\epsilon\omega u_0} \Big/ \frac{2c_0^3 \rho_0}{b\omega^2} = \frac{1}{\beta\omega u_0} \Big/ \frac{1}{\delta\omega^2} . \qquad (3.12)$$

It is evident that the quantity Γ estimates the relative contribution of nonlinear and dissipative effects into the wave profile distortion. For $\Gamma \ll 1$ nonlinearity dominates, whereas for $\Gamma \gg 1$ dissipation prevails. A consistent derivation of the Burgers equation (3.8) from the equations of hydrodynamics is offered in [4].

Problem 3.3

Assuming that the sound absorption coefficient in water is defined by the value $\delta = 0.6 \cdot 10^{-17}$ s^2/cm and that it in air amounts to $\delta = 0.5 \cdot 10^{-14}$ s^2/cm, estimate the acoustic Reynolds number in problems 2.7, 2.8, 2.10.

Answer a) Re ≈ 22, b) Re ≈ 13, c) Re ≈ 300,

Problem 3.4

Let $\Pi(x, \tau)$ be a certain known solution to the Burgers equation (3.8) that corresponds to the boundary condition $\Pi(x = 0, \tau) = \Pi_0(\tau)$. Find a solution conforming to the superposition of the constant flow with velocity $V_0 = constant$ on the reference perturbation Π_0, i. e.

$$u(x = 0, \tau) = V_0 + \Pi_0(\tau) \ . \tag{3.13}$$

Answer This solution is given by formula

$$u(x, \tau) = V_0 + \Pi_0(x, \tau + \beta V_0 x) \ . \tag{3.14}$$

The propagation speed of the wave Π downstream is $\Delta c \approx \beta c_0^2 V_0 = \epsilon V_0$ higher than that in an unperturbed medium.

Problem 3.5

Find a stationary solution of the Burgers equation satisfying the conditions of a symmetric jump $u(\tau \to -\infty) = -u_0$ and $u(\tau \to \infty) = u_0$. Employ the transformation (3.14) in the previous problem to construct a stationary solution which satisfies the conditions $u(\tau \to -\infty) = u_1$ and $u(\tau \to \infty) = u_2 > u_1$.

Answer The stationary wave is sought in the form $u(x, \tau) = u(\tau + Cx)$ where the constant C is defined from the conditions as $\tau \to \pm\infty$. In the first case the stationary solution has the form

$$u = u(\tau) = u_0 \tanh(\beta u_0 \tau / 2\delta) = u_0 \tanh(\omega \tau / 2\Gamma) \ , \tag{3.15}$$

which describes a symmetric shock wave travelling with the sound velocity. The front thickness is inversely proportional to the jump value u_0. Assuming $V_0 = (u_1 + u_2)/2$, $u_0 = (u_2 - u_1)/2$ from (3.13) and (3.14) we obtain for the moving shock front

$$u = \frac{u_1 + u_2}{2} + \frac{u_2 - u_1}{2} \tanh[\frac{\beta u_0}{2\delta}(\tau + \beta \frac{u_1 + u_2}{2} x)] \ . \tag{3.16}$$

It seems useful to make sure that the travel velocity of the weak shock wave front (3.16) does not depend on its thickness and coincides with the velocity (2.23) of the discontinuity propagation.

Problem 3.6

Demonstrate that the Burgers equation, through the change of the variables

$$u = \frac{\partial S}{\partial \tau} \ , \qquad S = \frac{2\delta}{\beta} \ln U \tag{3.17}$$

or

$$u = \frac{2\delta}{\beta} \frac{\partial \ln U}{\partial \tau} \ , \tag{3.18}$$

(Hopf-Cole substitution) is reduced to the linear diffusion equation. Find the general solution of the Burgers equation.

Solution From equation (3.8), obtain the equation for S

$$\frac{\partial S}{\partial x} - \frac{\beta}{2}(\frac{\partial S}{\partial \tau})^2 = \delta \frac{\partial^2 S}{\partial \tau^2} \ , \tag{3.19}$$

which after the transition of (3.17) to U is reduced to the linear parabolic equation

$$\frac{\partial U}{\partial x} = \delta \frac{\partial^2 U}{\partial \tau^2} \ , \tag{3.20}$$

coinciding in form with (3.2). The solution of this equation subject to the boundary condition $U(x = 0, t) = U_0(t)$ will be formulated similar to (3.3)

$$U(x, \tau) = \frac{1}{\sqrt{4\pi\delta x}} \int_{-\infty}^{\infty} U_0(t) \exp[-\frac{(t - \tau)^2}{4\delta x}] \, dt \ . \tag{3.21}$$

Taking the change (3.17) into account

$$U_0(t) = \exp[\frac{\beta}{2\delta}S_0(t)] , \qquad S_0(t) = \int^t u_0(t')\, dt' , \qquad (3.22)$$

the chain of transformations (3.22) $\rightarrow$ (3.21) $\rightarrow$ (3.18) provides the general solution to the Burgers equation, i.e. furnishes the expression of the field $u(x,\tau)$ in an arbitrary section x through the reference field $u_0(\tau)$. Let us show one more form in which the general solution can be written. Using (3.18) we obtain from (3.21), (3.22)

$$u(x,\tau) = \frac{\int_{-\infty}^{\infty} \frac{t-\tau}{\beta x} \exp[\frac{1}{2\delta}F(\tau,t,x)]\, dt}{\int_{-\infty}^{\infty} \exp[\frac{1}{2\delta}F(\tau,t,x)]\, dt} , \qquad (3.23)$$

where

$$F = \beta S_0(t) - \frac{(t-\tau)^2}{2x} , \qquad S_0(t) = \int^t u_0(t')\, dt' . \qquad (3.24)$$

Problem 3.7

On the basis of the general solution of the Burgers equation arrived at in the previous problem, consider the evolution of a harmonic reference signal $u_0(t) = a\sin(\omega t)$. Investigate its asymptotic behaviour as $x \to \infty$.

Answer Making use of the expansion [22)

$$\exp(z\cos\theta) = I_0(z) + 2\sum_{n=1}^{\infty} I_n(z)\cos(n\theta) , \qquad (3.25)$$

where I_n are modified Bessel functions, one can obtain from (3.18), (3.21) and (3.22)

$$u(x,\tau) = \frac{2a}{\mathrm{Re}} \frac{\sum_{n=1}^{\infty} n(-1)^{n+1} I_n(\mathrm{Re}) \exp(-\delta n^2\omega^2 x)\sin(n\omega\tau)}{I_0(\mathrm{Re}) + 2\sum_{n=1}^{\infty}(-1)^n I_n(\mathrm{Re}) \exp(-\delta n^2\omega^2 x)\cos(n\omega\tau)} .$$

$$(3.26)$$

Here the parameter combination $a\beta/2\omega\delta$ has the meaning of the acoustic Reynolds number (3.11). For $\delta\omega^2 x \gg 1$ the exponents in (3.26) drop markedly with n resulting in that the first harmonic is the only one remaining

$$u(x,\tau) \approx \frac{2a}{\mathrm{Re}} \cdot \frac{I_1(\mathrm{Re})}{I_0(\mathrm{Re})} \exp(-\delta\omega^2 x)\sin(\omega\tau) \ . \tag{3.27}$$

For small and large Reynolds numbers, using the asymptotic Bessel functions, a harmonic damping as per the linear acoustics laws can be formulated

$$u(x,\tau) \approx \exp(-\delta\omega^2 x)\sin(\omega\tau)\begin{cases} a & \mathrm{Re} \ll 1 \\ 4\delta\omega/\beta & \mathrm{Re} \gg 1 \end{cases} \ . \tag{3.28}$$

In the latter case the harmonic amplitude does not depend on its reference value.

Problem 3.8

Employing the general solution of the Burgers equation, consider the unipolar pulse evolution by approximating it as a δ-function at the input: $u_0 = A\delta(t)$. Introduce the Reynolds number for this problem and discuss the limiting cases $\mathrm{Re} \ll 1$ and $\mathrm{Re} \gg 1$.

Answer The solution has the form

$$u(x,\tau) = \frac{2\delta}{\beta} \frac{1}{\sqrt{4\pi\delta x}} \frac{(\mathrm{e}^{\mathrm{Re}} - 1)\mathrm{e}^{-\tau^2/(4\delta x)}}{1 + \frac{1}{2}(\mathrm{e}^{\mathrm{Re}} - 1)(1 + \Phi(\tau/\sqrt{4\delta x}))} \ , \tag{3.29}$$

where $\Phi(z) = (2/\sqrt{\pi}) \int_0^z \exp(-t^2)\,dt$ is the error integral and $\mathrm{Re} = A\beta/(2\delta)$. For $\mathrm{Re} \ll 1$ the result in (3.29) coincides with the linear solution (3.6). For $\mathrm{Re} \gg 1$ equation (3.29) implies that the pulse has a universal triangular form

$$u(x,\tau) \approx \begin{cases} -\tau/(\beta x) \ , & -T < \tau < 0 \\ 0 \ , & \tau < -T \ , \tau > 0 \end{cases} \tag{3.30}$$

where $T = \sqrt{2A\beta x}$ is the pulse duration. To derive formula (3.30) one has to use the asymptotics of the function $\Phi(-z)$ as $z \to \infty$.

Problem 3.9

Let $\Pi(x, \tau)$ be a certain known solution of the Burgers equation satisfying the boundary condition $\Pi(x = 0, t) = \Pi_0(t)$. Analyze the interaction of this wave with the linear profile of the flow (see Problem 1.9) on the basis of the general form of the Burgers equation solution (see Problem 3.6) for the boundary condition

$$u(x = 0, t) = \gamma t + \Pi_0(t) \ . \tag{3.31}$$

Examine the cases of $\gamma > 0$ and $\gamma < 0$.

Answer Nonlinear interaction with the linear profile brings about the change of the characteristic amplitude and frequency as well as of the wave $\Pi(x, \tau)$ evolution rate. The solution can be written as

$$u(x, \tau) = \frac{\gamma \tau}{1 - \beta \gamma x} + \frac{1}{1 - \beta \gamma x} \Pi(\frac{x}{1 - \beta \gamma x}, \frac{\tau}{1 - \beta \gamma x}) \ . \tag{3.32}$$

For $\gamma > 0$ and $\beta \gamma x \to 1$ the characteristic amplitude and frequency of the wave increase without bound.

Problem 3.10

Using the saddle-point method, find an asymptotic solution to the Burgers equation (3.8) for large Reynolds numbers ($\delta \to 0$). Interpret this solution graphically.

Solution The formula for the general solution (3.23) of the Burgers equation incorporates integrals of the type

$$I = \int_{-\infty}^{+\infty} f(t) \exp(\frac{1}{2\delta} F(\tau, t, x)) \, dt \ . \tag{3.33}$$

As $\delta \to 0$ the main contribution to the integral will be from the neighbourhoods of the points in which the function F has its maximum. Let t_K be one of these points, which can be found from the equation

$$\frac{\partial F}{\partial t} = 0 \ , \qquad 0 = \frac{t_K - \tau}{x} - \beta u_0(t_K) \ . \tag{3.34}$$

In the neighbourhood of this point the function F can be expanded into a series confined to quadratic terms

$$F(\tau, t, x) \approx F_K + F_K''(t - t_K)^2/2 , \qquad (3.35)$$

where $F_K = F(\tau, t_K, x)$, $\quad F'' = x^{-1} - \beta u_0'(t_K) < 0$. Then the integral (3.33) can be rendered as a sum of contributions in the saddle-points

$$I = \sum I_K , \qquad I_K = f(t_K)\sqrt{\frac{4\pi\delta}{|F_K''|}} \exp(\frac{F_K}{2\delta}) . \qquad (3.36)$$

As $\delta \to 0$ one term in this sum will be prevailing, which corresponds to the absolute maximum of function F. Therewith, the general solution (3.23) yields the asymptotic result

$$u(x, \tau) = \frac{t_*(x, \tau) - \tau}{\beta x} , \qquad (3.37)$$

where $t_*(x, \tau)$ is the coordinate of the absolute maximum of the function

$$F(\tau, t, x) = \beta S_0(t) - \frac{(t - \tau)^2}{2x} , \qquad S_0(t) = \int^t u_0(t') \, dt' . \quad (3.38)$$

The absolute maximum seeking procedure lends itself well to clear graphic representation. It is evident that the coordinate $t_*(x, \tau)$ is the first point in which a mobile straight line h descending from infinity in parallel with the abscissa axis t touches the function F. It seems more convenient, however, to consider the first touching point between the function $\beta S_0(t)$ and the parabola

$$\alpha(\tau, t, x) = h + \frac{(t - \tau)^2}{2x} , \qquad (3.39)$$

descending (as h decreases) on the function $\beta S_0(t)$ (see Figure 3.1).

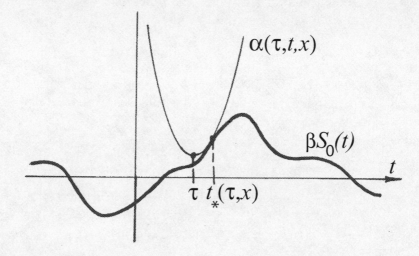

Figure 3.1: *Graphical interpretation of the lowering of parabola as a method for finding the absolute maximum of the function (3.38).*

Problem 3.11

Using the asymptotic solution of the Burgers equation obtained in the previous problem, analyze the evolution of a unipolar pulse by approximating it with a δ-function at the input: $u_0(t) = A\delta(t)$.

Solution For the function $\beta S_0(t)$ defined by formula (3.38) we have $\beta S_0 = \beta A\Theta(t)$ where $\Theta(t)$ is the Heaviside function. A graphic procedure for finding the absolute maximum coordinate is for this case depicted in Figure 3.2.

Fix the distance x, i.e. the parabola width (3.39). If $\tau > 0$, then the parabola will evidently touch the step by its centre $t = \tau$, i. e. $t_*(x, \tau) = \tau$; in this case, according to (3.37) the field $u(x, \tau) \equiv 0$ for all $\tau > 0$. For $\tau < 0$ one critical parabola α_* exists which touches $\beta S_0(t)$ simultaneously in the two points $t = 0$ and $t = -T$. It is clear that then $h = 0$ and the position α_* is revealed from the set of

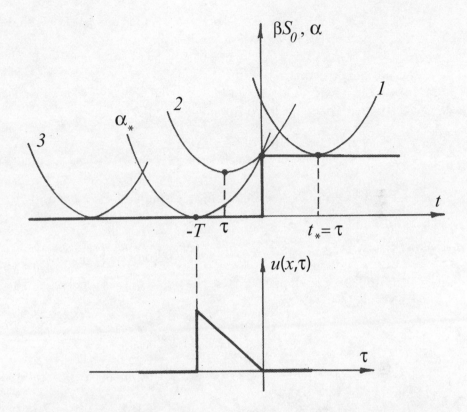

Figure 3.2: *Top: the graphical procedure for finding the coordinate of absolute maximum for an initial delta-shaped pulse. Below: the asymptotic triangular shape of that pulse at large distances and large Reynolds numbers.*

equations

$$\alpha_*(\tau, -T, x) = \frac{(-T-\tau)^2}{2x} = 0 \ , \qquad \alpha_*(\tau, 0, x) = \frac{\tau^2}{2x} = \beta A \ .$$
(3.40)

Hence it follows that the coordinate of the vertex of this critical parabola is equal to

$$-T = -(2\beta A x)^{1/2} \ .$$
(3.41)

Assuming that $-T < \tau < 0$, it can be easily seen that parabola 2 in

Figure 3.2 will touch $\beta S_0(t)$ at the point $t_* = 0$. From (3.37) we find that $u(x, \tau) = -\tau/(\beta x)$. At last, assuming that $\tau < -T$, i.e. shifting the centre of the mobile parabola 3 to the left of the centre of the critical parabola α_* we again find that $t_* = \tau$, $u \equiv 0$.

From what is said above, it can be seen that the asymptotic profile has a triangular shape for large Reynolds numbers

$$u(x, \tau) = 0 \ , \quad (\tau < -T, \tau > 0) \ ;$$
$$u(x, \tau) = -\tau/(\beta x) \ , \quad (-T < \tau < 0) \ . \tag{3.42}$$

The pulse duration $T(x)$ determined from formula (3.41) and the peak perturbation value $u_{max}(x)$ are equal to

$$T(x) = (2\beta A x)^{1/2} \ ; \quad u_{max}(x) = u(x, \tau = -T) = (2A/(\beta x))^{1/2} \tag{3.43}$$

(see also Problem 3.8). It is clear that the pulse area $u_{max}(x)T(x)/2 = A = \text{constant}$.

Problem 3.12

Making use of the graphic procedure in 3.10, analyze the process of two unipolar pulses' interaction

$$u_0(t) = A_1\delta(t) + A_2\delta(t - t_0) \ . \tag{3.44}$$

for large Reynolds numbers. Find the asymptotic wave form resulting from the merging of the pulses.

Answer The critical parabolas α_* (see the previous problem) and the corresponding pulse profiles are depicted in Figure 3.3.

It has to be reminded that the parabolas broaden with distance x. The discontinuity coordinates are readily obtained from the condition of double touching of function $\beta S_0(t)$ by parabola α (Figure 3.3 a,b). The distance x at which the discontinuity merging is observed is found from the condition of triple touching between α and $\beta S_0(t)$ (Figure 3.3 c). The asymptotic wave shape is a solitary triangular pulse (3.42) with duration $T = [2\beta(A_1 + A_2)x]^{1/2}$.

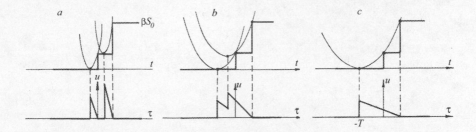

Figure 3.3: *Analysis of two compression pulses' interaction by means of the method of lowering of parabola.*

Problem 3.13

Under the conditions specified for the previous problem, consider the interaction of two δ-pulses of different polarity. Treat the case of $|A_1| = |A_2|$ separately.

Problem 3.14

Improve the solution of the problem (2.17) for a period of a saw-tooth wave taking into account that in a dissipative medium for large Reynolds numbers the shock front has a small but finite thickness and is described by formula (3.15).

Solution One needs to replace the step function $\mathrm{sgn}(\tau)$ with $\tanh[\beta u_S(x)\tau/(2\delta)]$. In the hyperbolic tangent argument we must take into consideration that the discontinuity value drops due to nonlinear damping as $u_S(z)/u_0 = \pi/(1+z)$ (see Problem 2.13); an appropriate growth is observed for the front thickness. Therefore, the formula (2.17) in Problem 2.14 will acquire the form

$$\frac{u}{u_0} = \frac{1}{1+z}[-\omega\tau + \pi\tanh(\frac{\pi}{1+z}\frac{\beta u_0}{2\delta}\tau)] \ ; \ \ -\pi < \omega\tau < \pi \ , \ \ \mathrm{Re} \gg 1 \ .$$
(3.45)

This is the Khokhlov solution. Substituting it into the Burgers equation (3.8) one can see that it is an exact solution to the equation.

Problem 3.15

Expand the Khokhlov solution into a Fourier series, calculate the harmonic amplitudes and analyze their behaviour at large distances.

Answer The series expansion (the Fay solution) has the form

$$\frac{u}{u_0} = \sum_{n=1}^{\infty} 2(\frac{\omega\delta}{u_0\beta})\{\sinh[n(1+z)(\frac{\omega\delta}{u_0\beta})]\}^{-1}\sin(n\omega\tau) . \qquad (3.46)$$

It offers a good description of the harmonic (at input $z = 0$) wave spectrum for large Re in the region where the front is stabilized, i.e. the nonlinear steepening and the dissipative flattening of the profile balance each other. The harmonic amplitude for $[\omega\delta/(u_0\beta)]z > 1$ in the Fay solution is approximately decreasing following the law $\exp(-n\delta\omega^2 x)$, i.e. slower than by the linear theory $\exp(-n^2\delta\omega^2 x)$. This is associated with the energy pumping from the lower harmonics to the higher ones. At distances $[\omega\delta/(u_0\beta)]z >\approx 2$ or $\delta\omega^2 x >\approx 2$ the main term in the Fay solution becomes the first one in the series (3.46) and the wave takes on the form

$$u = \frac{4\delta\omega}{\beta}e^{-\delta\omega^2 x}\sin(\omega\tau) = u_{max}(x)\sin(\omega\tau) . \qquad (3.47)$$

Formula (3.47) coincides with (3.28) and describes the nonlinear saturation effect: no matter how large the wave amplitude u_0 is when entering the nonlinear medium, at distances more than two lengths of linear damping $x >\approx 2/(\delta\omega^2) = 2x_d$ the wave amplitude can never exceeed

$$u_{max} = \frac{4\delta\omega}{\beta}e^{-\delta\omega^2 x} = \frac{2b\omega}{\epsilon c_0 \rho_0}\exp(-\frac{b\omega^2}{2c_0^3\rho_0}x) . \qquad (3.48)$$

Problem 3.16

Under the conditions of the Problems a) 2.7, and b) 2.8; estimate the dissipative length $x_d = 1/(\delta\omega^2) = 2c_0^3\rho_0/(b\omega^2)$ and find the maximal intensity of a wave which is transmitted over a distance $2x_d$. Assume for water that $\delta = 0.6 \cdot 10^{-17}$ s^2/cm.

Answer a) $x_d \approx 42$ m, $I_{max} \approx 10^{-4}$ W/cm^2, b) $x_d \approx 1$ km, $I_{max} \approx 4 \cdot 10^{-6}$ W/cm^2,

Chapter 4

SPHERICAL AND CYLINDRICAL WAVES AND NONLINEAR BEAMS

Problem 4.1

Consider converging, spherically-symmetric waves in the linear approximation. The reference shape of perturbation $u_0(t)$ is specified on the spherical surface of radius $r_0 \gg \lambda$ (where λ is a characteristic wave length). Using the method of slowly varying profile (Problem 1.5), simplify the linear wave equation

$$\Delta u - \frac{1}{c_0^2} \cdot \frac{\partial^2 u}{\partial t^2} = 0, , \qquad \Delta u = \frac{\partial^2 u}{\partial r^2} + \frac{2}{r} \cdot \frac{\partial u}{\partial r} . \qquad (4.1)$$

Solution Passing to the accompanying coordinate system $\tau = t + (r - r_0)/c_0$, $r_1 = \mu r$ and neglecting the small term $\sim \mu^2$ one obtains

$$\frac{\partial^2 u}{\partial \tau \partial r} + \frac{1}{r} \frac{\partial u}{\partial \tau} + \frac{c_0}{r} \frac{\partial u}{\partial r} = 0 . \qquad (4.2)$$

The ratio of the third term to the first one in equation (4.2) is a quantity on the order of $c_0/(r\omega_0) \sim \lambda/r$. Consequently, the third term is small everywhere except in the immediate neighbourhood of the focus $r = 0$. Neglecting the third term in (4.2) a simplified equation is obtained:

$$\frac{\partial u}{\partial r} + \frac{u}{r} = 0 . \qquad (4.3)$$

Its converging wave solution for r decreasing from r_0 to 0 is

$$u(r, t) = \frac{1}{r} u_0(\tau = t + \frac{r - r_0}{c_0}) . \qquad (4.4)$$

It increases infinitely as the wave converges to the focal point $r = 0$.

Problem 4.2

Derive the modified Burgers' equation (see (3.8)) using a generalisation of the simplified equation (4.3) by analogy with the method used in Problem 3.2. Assume both nonlinear and dissipative effects to be weak. As a result the wave profile is slowly distorted during the wave propagation.

Answer

$$\frac{\partial u}{\partial r} + \frac{u}{r} - \beta u \frac{\partial u}{\partial \tau} + \delta \frac{\partial^2 u}{\partial \tau^2} = 0 \ . \tag{4.5}$$

Here $\beta = \epsilon/c_0^2$, $\delta = b/(2c_0^3\rho_0)$, as in Problem 3.2.

Problem 4.3

Transform the modified Burgers equation (4.5) for spherical waves using dimensionless variables

$$U = -\frac{u}{u_0} \frac{r}{r_0} \ , \qquad \theta = \omega\tau \ , \qquad \xi = \beta\omega u_0 r_0 \ln(r_0/r) \ . \tag{4.6}$$

Comparing the obtained result with (3.10), determine the formal analogy between convergence and coordinate-dependent dissipation.

Answer

$$\frac{\partial U}{\partial \xi} = U \frac{\partial U}{\partial \theta} + \Gamma \exp(-\xi/z_0) \frac{\partial^2 U}{\partial \theta^2} \tag{4.7}$$

Here $\Gamma = \delta\omega/(\beta u_0)$ is the inverse Reynolds number (see (3.11), $z_0 = \beta\omega u_0 r_0$ is the dimensionless initial radius of the wave front. One can see, that Burgers' equation in the form (4.7) reduces the problem for the spherical wave in a homogeneous medium to a plane wave propagating in a medium with dissipation decaying exponentially with distance (ξ increasing from 0 to ∞).

Problem 4.4

Derive the modified Burgers' equation for a cylindrically converging wave, by analogy with Problems 4.1 and 4.2.

Answer

$$\frac{\partial u}{\partial r} + \frac{u}{2r} - \beta u \frac{\partial u}{\partial \tau} + \delta \frac{\partial^2 u}{\partial \tau^2} = 0 \ . \tag{4.8}$$

Here we use the same notation as in (4.5).

Problem 4.5

Transform equation (4.8) using the change of variables

$$U = -\frac{u}{u_0}\sqrt{\frac{r}{r_0}} \ , \qquad \theta = \omega\tau \ , \qquad \xi = 2\beta\omega u_0 r_0 (1 - \sqrt{r/r_0}) \ . \tag{4.9}$$

Show the meaning of the obtained equation (like in Problem 4.3).

Answer

$$\frac{\partial U}{\partial \xi} = U \frac{\partial U}{\partial \theta} + \Gamma(1 - \frac{\xi}{2z_0})\frac{\partial^2 U}{\partial \theta^2} \ . \tag{4.10}$$

It is evident that equation (4.10) is equivalent to the Burgers equation for the plane waves in a medium where dissipative characteristics diminish according to a linear law as r changes from r_0 to 0 (in this case ξ increases from zero to $2z_0$ where $z_0 = \beta\omega u_0 r_0$ is a dimensionless reference radius of the front).

Problem 4.6

Find the distance that must be covered by the reference harmonic spherically symmetric wave in a nondissipative medium to induce a discontinuity formation in the wave profile. Consider a) converging and b) diverging waves.

Solution Inasmuch as for $\Gamma = 0$ equation (4.7) coincides with an ordinary equation of simple waves the coordinate of the discontinuity formation r_S in the initial harmonic wave has to be derived from the condition $\xi = \beta \omega u_0 r_0 |\ln(r_0/r_S)| = 1$ (see Problem 2.4). Here the case $r_0 < r_S < \infty$ corresponds to the diverging wave, whereas $0 < r_S < r_0$ is for the wave converging to focus $r = 0$. The distance $|r_S - r_0|$ that is to be covered by the wave to become discontinuous amounts to

$$a) \quad |r_S - r_0| = r_0[1 - \exp(-\frac{1}{\beta \omega u_0 r_0})] \;, \qquad (4.11)$$

$$b) \quad |r_S - r_0| = r_0[\exp(\frac{1}{\beta \omega u_0 r_0}) - 1] \;, \qquad (4.12)$$

It is clear that for the converging waves (4.11) the inequality $|r_S - r_0| < (\beta \omega u_0)^{-1}$ is satisfied, i.e. a discontinuity is formed at shorter distances than for a plane wave. On the contrary, formula (4.12) implies that $|r_S - r_0| > (\beta \omega u_0)^{-1}$, i.e. a diverging wave has to cover a larger distance in order to reach a discontinuous state. The change of the nonlinear distortion build-up rate is related to the fact that in converging spherical waves the amplitude increases as r falls off (from r_0 to 0) whereas in diverging waves it decreases as r grows from r_0 to ∞.

Problem 4.7

Determine if a discontinuity can always be formed in a converging initially harmonic wave propagating in a nondissipative medium.

Solution In a cylindrically converging wave the condition for discontinuity formation, according to (4.9) has the form

$$\xi = 2\beta \omega u_0 r_0 (1 - \sqrt{r/r_0}) = 1 \;. \qquad (4.13)$$

Since $0 < r < r_0$, the maximal value ξ is attained for $r = 0$ amounting to $2\beta \omega u_0 r_0$. Provided the parameters on the radiating cylindrical surface are chosen such that $\beta \omega u_0 r_0 < 1/2$, the condition (4.13) is not fulfilled for any r and no discontinuity is formed when converging to the focus $r = 0$.

For the spherical wave condition, equation (4.13) has the form (see (4.6))

$$\xi = \beta \omega u_0 r_0 \ln(r_0/r) = 1 \; . \tag{4.14}$$

Here we are faced with the opposite case: no matter how small the combination of parameters $\beta \omega u_0 r_0$ is, on the radiating surface a small r can always be found near the focus $r = 0$ such that a discontinuity will form.

Problem 4.8

Generalize the Khokhlov solution (3.45) to spherical waves and analyze the shock front formation process in a converging wave taking dissipation into account.

Solution Using the notations in Problem 4.3 and making a comparison of an ordinary (3.10) and a spherical (4.7) Burgers equation lead to an expression for the profile of a spherical converging wave period

$$U = \frac{1}{1+\xi}[-\theta + \pi \tanh(\frac{\pi}{1+\xi}\frac{\theta}{2\Gamma}e^{\xi/z_0})] \; . \tag{4.15}$$

The shock front thickness is determined from the hyperbolic tangent argument

$$\Delta\theta_\Phi = \frac{2\Gamma}{\pi}(1+\xi)\exp(-\xi/z_0) = \frac{2\Gamma}{\pi}(1+z_0\ln(r_0/r))r/r_0 \; . \tag{4.16}$$

From analysis of expressions (4.16), it follows that the function $\Delta\theta_\Phi(r)$ has its maximum if the condition $z_0 = \beta\omega u_0 r_0 > 1$ is valid. This means that for $z_0 > 1$ double shock formations are observed. An initially narrow front broadens due to dissipation. Its thickness attains the maximum value $(2\Gamma/\pi)z_0 \exp(1/z_0 - 1)$ at the point $r = r_0 \exp(1/z_0 - 1)$. Then the nonlinear action becomes more pronounced again, and the front thickness tends to zero as the wave converges at its focus [10].

Problem 4.9

Using the quasioptical approximation of the theory of diffraction and the method of slowly varying profile (Problem 1.5), deduce a simplified equation for the beam in the linear approximation.

Solution Let us proceed from the linear wave equation written in Cartesian coordinates

$$\frac{\partial^2 u}{\partial x^2} + \frac{\partial^2 u}{\partial y^2} + \frac{\partial^2 u}{\partial z^2} - \frac{1}{c_0^2}\frac{\partial^2 u}{\partial t^2} = 0 \ . \tag{4.17}$$

Let the wave propagate along the beam axis x. In the quasioptical approximation a harmonic signal is usually considered. In this case the wave amplitude is assumed to change slowly both along the $x (\sim \mu x)$ axis and across the beam $(\sim \sqrt{\mu}y, \sim \sqrt{\mu}z)$

$$u = \exp(-i\omega t + i\omega x/c_0) \cdot A(x_1 = \mu x, y_1 = \sqrt{\mu}y, z_1 = \sqrt{\mu}z) \ . \tag{4.18}$$

If broadband signals or nonlinear propagation - where the signal spectrum is enriched with harmonics - are considered, the wave can not be treated as harmonic. Both its profile and spectrum are assumed to be slowly changing during propagation and formula (4.18) should be generalized

$$u = u(\tau = t - x/c_0, x_1 = \mu x, y_1 = \sqrt{\mu}y, z_1 = \sqrt{\mu}z) \ . \tag{4.19}$$

Substitute (4.19) into (4.17). The terms of the order of μ^0 cancel each other and the μ^2 terms are disregarded. As a result all remaining terms are of the same infinitesimal order μ^1. These terms form the simplified equation

$$\frac{\partial^2 u}{\partial x \partial \tau} = \frac{c_0}{2}\Delta_\perp u \ , \qquad \Delta_\perp = \frac{\partial^2}{\partial y^2} + \frac{\partial^2}{\partial z^2} \ . \tag{4.20}$$

For harmonic signals $u = A\exp(-i\omega\tau)$, (4.20) yields the familiar parabolic equation diffraction theory

$$-2i\kappa\frac{\partial A}{\partial x} = \Delta_\perp A \ , \qquad \kappa = \omega/c_0 \ . \tag{4.21}$$

Problem 4.10

Using the method of the previous problem, deduce the simplified Khokhlov-Zabolotskaya equation from the nonlinear wave equation

$$\Delta u - \frac{1}{c_0^2} \cdot \frac{\partial^2 u}{\partial t^2} = -\frac{\epsilon}{c_0^3} \cdot \frac{\partial^2 u^2}{\partial t^2} \ . \tag{4.22}$$

Solution On the assumption of the slow change of the wave profile and the beam shape (4.19) one obtains

$$\frac{\partial}{\partial \tau}\left(\frac{\partial u}{\partial x} - \frac{\epsilon}{c_0^2}u\frac{\partial u}{\partial \tau}\right) = \frac{c_0}{2}\Delta_\perp u \ . \tag{4.23}$$

This is the Khokhlov-Zabolotskaya equation. If the transverse coordinate dependence is neglected ($\Delta_\perp u = 0$) (4.23) will be transformed into the simple wave equation (1.13). If nonlinearity is disregarded ($\epsilon = 0$) (4.23) will be transformed into the equation of the linear theory of diffraction (4.20). Equation (4.23) describes the wave accounting simultaneously for nonlinear and diffraction effects. A more rigorous derivation of (4.23) from the equations of hydrodynamics is discussed in [4,7].

Problem 4.11

Acting as in Problem 3.2, obtain the expression for the dimensionless complex of parameters, N, making it possible to assess the relative contribution of nonlinear and diffraction effects to the distortion of a wave.

Solution Let a signal at input $x = 0$ be described by the function

$$u(x = 0, t) = u_0\, f(\vec{r}/a)\Phi(\omega t) \ . \tag{4.24}$$

Here $\vec{r} = \{y, z\}$ are the coordinates with respect to the beam cross-section, a is characteristic beam width; and u_0 and ω are the characteristic amplitude and frequency. Taking (4.24) into account we switch to the dimensionless variables of the type (3.9)

$$V = u/u_0 \ , \qquad \theta = \omega\tau \ , \qquad z = \beta\omega u_0 x = x/x_S \ , \qquad \vec{R} = \vec{r}/a \ .$$
$$\text{(4.25)}$$

Equation (4.23) will be reduced to the form

$$\frac{\partial}{\partial\theta}\left(\frac{\partial V}{\partial z} - V\frac{\partial V}{\partial\theta}\right) = \frac{N}{4}\Delta_\perp V \ . \tag{4.26}$$

Here $\Delta_\perp$ is the Laplace operator with respect to the normalized coordinates $\vec{R}$. The only parameter entering equation (4.26) is the number

$$N = \frac{2c_0^3}{\epsilon\omega^2 a^2 u_0} = \frac{1}{2\pi^2\epsilon M}\left(\frac{\lambda}{a}\right)^2 \ . \tag{4.27}$$

One can write N as a ratio of nonlinear and diffraction lengths

$$N = x_S/x_D = \frac{c_0^3}{\epsilon\omega u_0}\bigg/\frac{\omega a^2}{2c_0} \ . \tag{4.28}$$

Hence it follows that for $N \ll 1$ nonlinearity prevails, whereas diffraction dominates for $N \gg 1$.

Problem 4.12

Calculate in the linear approximation the variation of characteristics of the circular harmonic Gaussian beam

$$u(x = 0, r, t) = u_0 \exp(-r^2/a^2)\sin(\omega t) \tag{4.29}$$

due to diffraction.

Solution For the beams with round cross-section, equation (4.20) will acquire the form

$$\frac{\partial^2 u}{\partial x\, \partial\tau} = \frac{c_0}{2}\left(\frac{\partial^2 u}{\partial r^2} + \frac{1}{r}\frac{\partial u}{\partial r}\right) \ . \tag{4.30}$$

A solution to (4.30) subject to the boundary condition (4.29) can be achieved via the method of separation of variables or by using integral transformations. Direct substitution can be used as well to check if the solution (4.30) has the form [6]

$$u \approx \frac{u_0}{\sqrt{1 + x^2/x_D^2}} \exp(-\frac{r^2}{a^2} \frac{1}{1 + x^2/x_D^2})$$

$$\sin(\omega\tau + \arctan(x/x_D) - \frac{r^2}{a^2} \frac{x/x_D}{1 + x^2/x_D^2} \quad , \tag{4.31}$$

where $x_D = \omega a^2/(2c_0)$ is a characteristic diffraction length. The solution (4.31) describes the transformation of the reference plane wave into a spherically diverging one. The beam axis amplitude falls off as

$$u_{max} = u_0 \left(1 + x^2/x_D^2\right)^{-1/2} . \tag{4.32}$$

For $x \gg x_D$ the amplitude decreases as $u_{max} \approx u_0 x_D/x$ following the $\sim x^{-1}$ law of a spherically diverging wave. The beam width grows as

$$a(x) = a(1 + x^2/x_D^2)^{1/2} . \tag{4.33}$$

For $x \gg x_D$ and $a(x) \approx ax/x_D$, i.e. the width increases linearly with x, and the whole sound field is localized within a cone with the apex angle $\Delta\theta \approx 2a(x)/x \approx 4c_0/(\omega a)$. In addition it has to be noted that the wave phase at the beam axis acquires a shift of $\arctan(x/x_D)$. This suggests that the wave propagation speed at the beam axis is somewhat higher than that of the plane wave at the same frequency.

As the angular frequency ω (4.29) grows, the diffraction process attenuates and all the mentioned events show up at larger distances.

Problem 4.13

Making use of the solution (4.31) of the previous problem, demonstrate that a broadband signal (pulse) changes its shape in the far field region ($x \gg x_D$). The diffraction leads to a profile shape differentiation on the beam axis.

Solution All of the reference signal harmonics are described by expression (4.31), which for $x \gg x_D$, $r = 0$ takes the form

$$u \approx u_0(\omega)\frac{x_D}{x} \sin(\omega\tau + \pi/2) = \frac{\omega a^2}{2c_0 x}u_0(\omega) \cos(\omega\tau) . \tag{4.34}$$

The signal form is defined by a sum of all harmonics (4.34)

$$u = \frac{a^2}{2c_0 x} \int_{-\infty}^{\infty} u_0(\omega) \omega \cos(\omega\tau) \, d\omega = \frac{a^2}{2c_0 x} \frac{\partial}{\partial \tau} \int_{-\infty}^{\infty} u_0(\omega) \sin(\omega\tau) \, d\omega \ .$$

$$(4.35)$$

The last integral is the reference form of the pulse

$$\int_{-\infty}^{\infty} u_0(\omega) \sin(\omega\tau) \, d\omega = u(x = 0, \tau) \ . \tag{4.36}$$

Comparing (4.36) and (4.35) we find

$$u(x \gg x_D, \tau) = \frac{a^2}{2c_0 x} \cdot \frac{\partial}{\partial \tau} u(x = 0, \tau) \ , \tag{4.37}$$

i.e. the far region signal lends itself to differentiation.

Problem 4.14

Show that the compression and rarefaction regions of the nonlinear diffracting wave are distorted to different extent so that the reference harmonic signal profile turns non-symmetric as it propagates. Make use of the fact that different harmonics appear out of phase with respect to each other.

Solution To obtain a comprehensive solution, let us formulate the wave profile approximately as a sum of the first and the second harmonics only

$$u = A_1(x) \sin[\omega\tau + \phi_1(x)] + A_2(x) \sin[2\omega\tau + \phi_2(x)] \ . \tag{4.38}$$

It is evident that the second harmonic has an amplitude A_2 which is small compared to A_1. As the harmonic frequency is higher, the diffraction phase shift ϕ_2 for this harmonic is less than ϕ_1 (see Problem 4.12). Taking these factors into account the graphic summation of the two sinusoids (4.38) will indeed yield a nonsymmetric profile (Figure 4.1).

The compression region is shortened and more sharp, whereas the rarefaction region is extended and more smooth. Harmonics interfere such that the positive peak value of the perturbation exceeds its reference ($x = 0$) value.

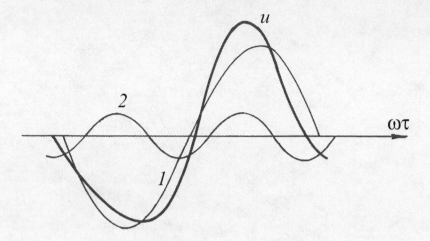

Figure 4.1: *Explanation of the asymmetry in the distortion of half-periods of compression and rarefaction of an initially harmonic signal. The asymmetry appears because of the different rates of diffraction of different harmonics. The resulting profile u is constructed as a sum of two thin curves for the first (1) and second (2) harmonics.*

Problem 4.15

Employing the model equation (4.22) and the method of successive approximations, calculate the amplitude of the difference frequency wave $\Omega = \omega_1 - \omega_2$ excited in a nonlinear medium under the interaction of two damped nondiffracting high frequency waves with close frequencies ω_1 and ω_2:

$$u^{(1)} = f(y,z)e^{-x/x_d}[A_1 e^{-i\omega_1(t-x/c_0)} + A_2 e^{-i\omega_2(t-x/c_0)}] + c.c. \ . \quad (4.39)$$

Here x_d is a characteristic damping length for ω_1 and ω_2, the function $f(y,z)$ describes the transverse structure of these waves' beam, and "c.c." means "complex conjugated" terms.

Solution Using (4.39) as the first approximation, obtain the following equation from (4.22) to find the second approximation

$$\Delta u^{(2)} - \frac{1}{c_0^2}\frac{\partial^2 u^{(2)}}{\partial t^2} = -\frac{\epsilon}{c_0^3}\cdot\frac{\partial^2 u^{(1)}{}^2}{\partial t^2} . \tag{4.40}$$

The right-hand side of (4.40) which describes nonlinear sources at the difference frequency Ω by taking (4.39) into account, will take the form

$$q(x,y,z)e^{-i\Omega t} + c.c. ;$$

$$q = \frac{2\epsilon}{c_0^3}\Omega^2 A_1 A_2 f^2(y,z)\exp[-\frac{2x}{x_d}+i\frac{\Omega}{c_0}x] + c.c. . \tag{4.41}$$

Seeking a solution of (4.40) in the form $u^{(2)} = A_- \exp(-i\Omega t)$ we obtain the inhomogeneous Helmholtz equation

$$\Delta A_- + K^2 A_- = q , \qquad K = \Omega/c_0 , \tag{4.42}$$

for the difference frequency wave amplitude A_-. The solution of this equation is written as

$$A_-(\vec{R}) = -\frac{1}{4\pi}\int_v q(\vec{R}_1)\frac{e^{iKr}}{r}\,dV_1 , \qquad r = |\vec{R} - \vec{R}_1| . \tag{4.43}$$

Here $\vec{R} = \{x,y,z\}$ is the radius vector of the observation point, and $\vec{R}_1 = \{x_1, y_1, z_1\}$ is the point inside the volume V_1 occupied by the ω_1 and ω_2 intersection region.

In the far diffraction region where $|\vec{R}| \gg |\vec{R}_1|$ it can be assumed that

$$r = \sqrt{R^2 - 2\vec{R}\vec{R}_1 + R_1^2} \approx$$
$$R(1 - \vec{R}\vec{R}_1/R^2) = R - (xx_1 + yy_1 + zz_1)/R . \tag{4.44}$$

Substituting expressions (4.41), (4.44) into the integral (4.43) we reduce the latter to the form

$$A_- \approx -\frac{\epsilon\Omega^2}{2\pi c_0^3}A_1 A_2 \frac{e^{iKR}}{R}D_t D_l . \tag{4.45}$$

Of special interest is the structure of expressions D_t, D_l defining the radiation directivity associated with the transverse

$$D_t = \int \int_{\infty}^{\infty} f^2(y,z) e^{-i\frac{K}{R}(yy_1 + zz_1)} \, dy_1 \, dz_1 \ , \qquad (4.46)$$

and the longitudinal (along the x axis)

$$D_l = \int_0^{\infty} \exp[-\frac{2x_1}{x_d} + iK(1 - x/R)x_1] \, dx_1 \ . \qquad (4.47)$$

distribution of the primary field $u^{(1)}$ (4.39). The integral (4.46) appears as an expansion into the angular spectrum of the function $f^2(y,z)$. It has the same form as that observed in the case when the wave Ω is radiated directly by a high frequency wave with a ω_1 and ω_2 source. The integral (4.47) describes the Ω directivity, with the wave being excited by spatially distributed nonlinear sources

$$D_l = [2/x_d - iK(1 - x/R)]^{-1} \ ,$$
$$|D_l| = \frac{x_d}{2}[1 + (Kx_d)^2 \sin^4(\theta/2)]^{1/2} \ . \qquad (4.48)$$

In formula (4.48) the relationship $1 - x/R = 1 - \cos\theta = 2\sin^2(\theta/2)$ was used; θ is the angle between the beam axis x and the direction with respect to the observation point.

For $K x_d \gg 1$ (many difference frequency wave lengths are present along the interaction region) the radiation proceeds at small angles with respect to the axis. The characteristic angular width of the directivity diagram is, as follows from (4.48), equal to

$$\Delta\theta_l \sim (Kx_d)^{-1/2} \sim (\Lambda/x_d)^{1/2} \ . \qquad (4.49)$$

The width (4.49) defined by the longitudinal distribution of the field $u^{(1)}$ is, as a rule, much less than the width defined by the transverse distribution (4.46), which is why formula (4.49) is responsible for the high directivity of the low-frequency signal.

Problem 4.16

Evaluate the angular width of the low-frequency radiation directivity defined by formula (4.49) of the previous problem. A 100 kHz difference frequency signal is excited by two pump waves with frequencies 1 MHz and 1.1 MHz in water. The absorption coefficient value is given in Problem 3.3.

Answer $\Delta\theta_l \sim 2 \cdot 10^{-2}$ is the angular width approaching one degree.

Problem 4.17

Calculate a longitudinal aperture factor (4.47) for the interaction range of nondecaying ($x_d \to \infty$) pump waves which is, for $x = l$, bounded by a filter absorbing high frequencies completely and only transmitting low frequency.

Solution

$$D_l = \int_0^\infty \exp[iK(1 - x/R)x_1]\,dx_1 = \frac{\exp[iKl(1 - x/R)] - 1}{iK(1 - x/R)} \ ,$$

$$|D_l| = l\mathrm{sinc}[Kl\sin^2(\theta/2)]) \ , \tag{4.50}$$

Here we made use of the following notation: $\mathrm{sinc}(x) \equiv \sin(x)/x$ and took into consideration that $1 - x/R \approx 2\sin^2(\theta/2)$. It has to be noted that the directivity diagram described by (4.50) contains lateral lobes (of course, it assumed that $Kl \gg 1$ i.e. many wave lengths fit inside the interaction length). The presence of lateral lobes is associated with the abrupt (down to zero) jump of the interacting wave amplitudes for $x = l$. In the case when the interaction range was bounded by an exponential damping law no lobes were observed (see (4.48)).

Chapter 5

HIGH INTENSITY ACOUSTIC NOISE

Problem 5.1

Neglecting frequency fluctuations, find the probability distribution and the average for the plane quasi-monochromatic wave discontinuity (shock) formation length assuming the probability distribution of the amplitude $W_a(a)$ is known.

Solution The results of Problem 2.4 imply that the discontinuity formation length x_S for a plane monochromatic wave is equal to $x_S = c_0^2/(\epsilon \omega a)$, where ω is frequency and a is the wave amplitude. The same formula can be applied for a quasimonochromatic wave as well, when the amplitude and the frequency change insignificantly over one wave period. Therefore, the problem is reduced to that of a nonlinear transformation

$$x_S = f(a) \ . \tag{5.1}$$

Provided the inverse function $a = f^{-1}(x_S)$ is single-valued, the probability distribution $W_x(x_S)$ is related with $W_a(a)$ through

$$W_x(x_S) = W_a[f^{-1}(x_S)]|df^{-1}(x_S)/dx_S| \ . \tag{5.2}$$

For the moments of the quantity x_S the following expression is valid

$$\langle x_S^n \rangle = \int_{-\infty}^{\infty} f^n(a)W(a)\,da \ . \tag{5.3}$$

For the probability distribution of the shock formation length and its average we get

$$W_x(x_S) = W_a(\frac{c_0^2}{\epsilon \omega x_S})(\frac{c_0^2}{\epsilon \omega x_S^2}) \tag{5.4}$$

$$\langle x_S \rangle = \frac{c_0^2}{\epsilon \omega} \int_0^{\infty} W_a(a)a^{-1}\,da \ . \tag{5.5}$$

Problem 5.2

Through the results of Problem 5.1, analyze the two different conditions:

(a) the amplitude of the signal is distributed uniformly inside the interval $[a_1, a_2]$;

(b) the input signal has Gaussian statistics with dispersion σ^2. Take into account the probability distribution of the amplitude of a Gaussian signal

$$W_a(a) = \frac{a}{\sigma^2} \exp(-\frac{a^2}{2\sigma^2}) \ . \tag{5.6}$$

which is also known as "Rayleigh distribution".

Answer Using (5.5) calculate probablility distributions and averages of shock formation distance:

a)

$$W_x(x_S) = \begin{cases} \frac{c_0^2}{(a_2 - a_1)\epsilon \omega x_S^2} \ , & x_S \subset [\frac{c_0^2}{\epsilon \omega a_2}, \frac{c_0^2}{\epsilon \omega a_1}] \ , \\ 0 & x_S \not\subset [\frac{c_0^2}{\epsilon \omega a_2}, \frac{c_0^2}{\epsilon \omega a_1}] \end{cases} \tag{5.7}$$

$$\langle x_S \rangle = \frac{c_0^2}{(a_2 - a_1)\epsilon \omega} \ln \frac{a_2}{a_1} \ . \tag{5.8}$$

b)

$$W_s(x_S) = \frac{c_0^4}{\epsilon^2 \omega^2 \sigma^2 x_S^3} \exp\{-\frac{c_0^4}{2\epsilon^2 \omega^2 \sigma^2 x_S^2}\} \ . \tag{5.9}$$

$$\langle x_S \rangle = \sqrt{\frac{\pi}{2}} \frac{c_0^2}{\epsilon \omega \sigma} \ . \tag{5.10}$$

Problem 5.3

Find the probability distribution of the quasi-monochromatic wave discontinuity (shock) amplitude

$$u = a \sin(\omega \tau + \phi) \ , \tag{5.11}$$

assuming the input signal to be Gaussian. The frequency fluctuations are to be disregarded.

Solution The discontinuity amplitude u_S is defined from the equation (see Problem 2.13)

$$\arcsin(\frac{u_S}{a}) = \frac{\epsilon}{c_0^2}\omega u_S x \ . \tag{5.12}$$

Using formula (5.2) and considering that for $a < c_0^2/(\epsilon\omega x)$ discontinuities would not form (formally their amplitude equals zero), we have

$$W_u(u_S) = \delta(u_S) \int_0^{c_0^2/(\epsilon\omega x)} W_a(a) \, da+$$

$$W_a[\frac{u_S}{\sin(\epsilon c_0^{-2}\omega u_S x)}]\frac{d}{du_S}[\frac{u_S}{\sin(\epsilon c_0^{-2}\omega u_S x)}] \tag{5.13}$$

for the probability distribution of the discontinuity amplitude $W_u(u_S)$. For a Gaussian input signal when the amplitude distribution follows the Rayleigh law (5.6) we obtain from (5.13)

$$W_u(u_S) = \delta(u_S)[1 - \exp(-c_0^4/2(\epsilon\omega\sigma x)^2)]+$$

$$\exp[-\frac{u_S^2}{2\sigma^2\sin^2(\epsilon c_0^{-2}\omega u_S x)}] \cdot \frac{1}{2\sigma^2}\frac{d}{du_S}[\frac{u_S}{\sin(\epsilon c_0^{-2}\omega u_S x)}]^2 \ . \tag{5.14}$$

Problem 5.4

Find the time averaged number of discontinuities $\bar{n}$ at the distance x measured from the input for a quasimonochromatic Gaussian input signal. Employ the results of the previous problem.

Answer

$$\bar{n} = \lim_{T\to\infty}\frac{N}{T} = \frac{\omega}{2\pi}\int_{c_0^2/(\epsilon\omega x)}^\infty W_a(a) \, da = \frac{\omega}{2\pi}e^{-1/(2z^2)} \ , \qquad z = \frac{\epsilon}{c_0^2}\omega\sigma x \ . \tag{5.15}$$

Problem 5.5

At the initial stage of the nonlinear manifestation (distances $x/x_S \ll 1$), the following expressions (see Problem 1.13) hold for the simple wave higher harmonic amplitudes

$$A_n(x) = a^n [n \frac{\epsilon}{c_0^2} \omega x]^{n-1}/(2^{n-1} \cdot n!) \ . \tag{5.16}$$

Assuming that a regular monochromatic signal of amplitude a_0 and a Gaussian quasimonochromatic signal with dispersion σ^2, both of the same intensity ($\sigma^2 = a_0^2/2$), are specified at the input. Compare the intensities of the higher harmonics of the noise signal ($\langle A_n^2 \rangle/2$) and the regular signal ($\overline{A}_n^2/2$).

Answer $\langle A_n^2 \rangle / \overline{A}_n^2 = n!$.

Problem 5.6

Consider an input signal to be stationary, Gaussian, with zero average, with the correlation function $B_0(\rho)$, and with spectrum $S_0(\omega)$ - find the correlation function and the simple wave spectrum at the initial stage restricting yourself to the first approximation in the solution of the simple wave equation by the method of perturbations.

Solution In the first approximation the solution to the simple wave equation (1.13) can be rendered as

$$u(x, \tau) = u_0(\tau) + u_1(x, \tau) \ , \tag{5.17}$$

$$u_1(x, \tau) = \frac{\epsilon}{2c_0^2} x \frac{d}{d\tau} y(\tau) \ , \qquad y(\tau) \equiv u_0^2(\tau) \ , \tag{5.18}$$

i.e. the relation between u_1 and u_0 is thought of as a sequence of a quadratic detector and a differentiating chain. For the correlation function $y(\tau)$ under the Gaussian input signal we have

$$K_y(\rho) = \langle y(\tau) \, y(\tau + \rho) \rangle = B_0^2(0) + 2B_0^2(\rho) \ . \tag{5.19}$$

Taking into account the relation between the process and its derivative for the simple wave correlation function we obtain

$$B_u(x, \rho) = \langle u(x, \tau)\, u(x, \tau + \rho)\rangle = B_0(\rho) - \frac{\epsilon^2 x^2}{2c_0^4} \cdot \frac{d^2}{d\rho^2} B_0^2(\rho) \ . \quad (5.20)$$

Inasmuch as the spectrum convolution corresponds to the correlation function to the 2nd power, formula (5.20) yields

$$g_u(x, \omega) = \frac{1}{2\pi} \int_{-\infty}^{+\infty} B_u(x, \rho)\, \mathrm{e}^{i\omega\rho}\, d\rho = S_0 + \frac{\epsilon^2 x^2 \omega^2}{2c_0^4} S_0(\omega) \otimes S_0(\omega) \ ,$$

$$(5.21)$$

$$S_0(\omega) \otimes S_0(\omega) \equiv \int_{-\infty}^{+\infty} S_0(\omega - \Omega) S_0(\Omega)\, d\Omega \ , \quad (5.22)$$

for the simple wave spectrum.

Problem 5.7

Find the simple wave spectrum at the initial stage for the Gaussian input signal in the cases of:
(a) broadband noise with correlation function $B_0(\rho) = \sigma^2 \exp[-\Delta^2 \rho^2/2]$
(b) narrow band noise $B_0(\rho) = \sigma^2 \exp[-\Delta^2 \rho^2/2] \cos(\omega_0 \rho)$, $(\omega_0 \gg \Delta)$.
Perform an analysis as to where new spectral components arise.

Answer Using the results of the previous problem yields

$$a) \quad S_u(x, \omega) = \frac{\sigma^2}{\sqrt{2\pi\Delta^2}} (\mathrm{e}^{-\omega^2/(2\Delta^2)} + \frac{\epsilon^2 \omega^2 \sigma^2 x^2}{2\sqrt{2} c_0^4} \mathrm{e}^{-\omega^2/(4\Delta^2)}) \ . \quad (5.23)$$

Nonlinear interaction brings about the spectrum broadening

$$b) \quad S_u(x, \omega) = \frac{\sigma^2}{2\sqrt{2\pi\Delta^2}} \{\mathrm{e}^{-(\omega-\omega_0)^2/(2\Delta^2)} + \mathrm{e}^{-(\omega+\omega_0)^2/(2\Delta^2)} +$$

$$\frac{\epsilon^2 \omega^2 \sigma^2 x^2}{4\sqrt{2}\, c_0^4} [2\mathrm{e}^{-\omega^2/(4\Delta^2)} + \mathrm{e}^{-(\omega-2\omega_0)^2/(4\Delta^2)} + \mathrm{e}^{-(\omega+2\omega_0)^2/(4\Delta^2)}]\} \ . \quad (5.24)$$

New spectral components emerge near zero and near to the double frequency $\omega = 2\omega_0$.

Problem 5.8

Find the time-averaged correlation function of a simple wave at the initial stage for a quasi-monochromatic input signal with Gaussian phase fluctuations

$$u_0(\tau) = a_0 \cos(\omega_0 \tau + \phi(\tau)) \ , \tag{5.25}$$

assuming the structural phase fluctuation function being known as

$$D_\phi(\rho) = \langle (\phi(\tau + \rho) - \phi(\tau))^2 \rangle \ . \tag{5.26}$$

Provide a qualitative description of the wave spectral composition.

Answer

$$K_u(x, \rho) = \frac{a_0^2}{2} \cos(\omega_0 \rho) e^{-D_\phi(\rho)/2} - \frac{\epsilon^2 a_0^4 x^2}{32 c_0^4} \frac{d^2}{d\rho^2} \cos(2\rho\omega_0) e^{-2D_\phi(\rho)} \ , \tag{5.27}$$

or, taking the phase fluctuation slowness into account

$$K_u(x, \rho) \approx \frac{a_0^2}{2} \cos(\omega_0 \rho) e^{-D_\phi(\rho)/2} + \frac{\epsilon^2 \omega_0^2 a_0^4 x^2}{8 c_0^4} \cos(2\rho\omega_0) e^{-2D_\phi(\rho)} \ . \tag{5.28}$$

Nonlinear interaction in this case results in the appearance of spectral components near the second harmonic.

Problem 5.9

Find under the conditions of Problem 5.8 the simple wave spectrum at the initial stage for a signal with restricted and small phase fluctuations $(D_\phi(\rho) = 2(\sigma_\phi^2 - B_\phi(\rho)), \ \sigma_\phi^2 \ll 1)$, assuming the spectrum $g_\phi(\omega)$ is known.

Answer

$$g_u(x, \omega) =$$

$$\frac{a_0^2}{4}[(1 - \sigma_\phi^2)(\delta(\omega - \omega_0) + \delta(\omega + \omega_0)) + g_\phi(\omega - \omega_0) + g_\phi(\omega + \omega_0)]$$

$$+\frac{\epsilon^2 \omega^2 a_0^4 x^2}{64 c_0^4}[(1 - 4\sigma_\phi^2)(\delta(\omega - 2\omega_0) + \delta(\omega + 2\omega_0))+$$

$$4g_\phi(\omega - 2\omega_0) + 4g_\phi(\omega + 2\omega_0)] \ . \tag{5.29}$$

Here the first term in square brackets describes the reference signal spectrum whereas the second term stands for the components near the second harmonic emerging due to nonlinear interaction. In front of the second term ω can be replaced with $2\omega_0$.

Problem 5.10

Under the same conditions as in Problem 5.8, find the simple wave spectrum at the initial stage for a signal with frequency fluctuations ($\Omega = \partial\phi/\partial t$) for the case of large and slow frequency fluctuations where the structural phase function can be approximated as $D_\phi(\tau) = \overline{\Omega^2}\tau^2$. Find the spectrum width $\Delta\omega_i$ at the first and second harmonics.

Answer

$$g_u(x, \omega) =$$

$$\frac{a_0^2}{4\sqrt{2\pi\overline{\Omega^2}}}\{\exp[-(\omega - \omega_0)^2/(2\overline{\Omega^2})] + \exp[-(\omega + \omega_0)^2/(2\overline{\Omega^2})]\}+$$

$$+\frac{\epsilon^2\omega^2 a_0^4 x^2}{64 c_0^4} \cdot \frac{1}{\sqrt{4\pi\overline{\Omega^2}}}\{\exp[-(\omega - 2\omega_0)^2/(4\overline{\Omega^2})]+$$

$$\exp[-(\omega + 2\omega_0)^2/(4\overline{\Omega^2})]\}$$

$$\Delta\omega_1 = \overline{\Omega} \ , \qquad \Delta\omega_2 = \sqrt{2}\overline{\Omega} \ . \tag{5.30}$$

Problem 5.11

Find the simple wave spectrum at the initial stage for an input signal with small amplitude fluctuations

$$u_0(\tau) = a_0(1 + \alpha(\tau)) \cos(\omega_0 \tau + \phi_0) , \qquad \langle \alpha^2 \rangle \ll 1 , \qquad (5.31)$$

assuming the spectrum of amplitude fluctuations $g_\alpha(\omega)$ to be known. Compare with the case of small phase fluctuations (Problem 5.9).

Answer

$$g_u(x, \omega) =$$

$$\frac{a_0^2}{4} [\delta(\omega - \omega_0) + \delta(\omega + \omega_0) + g_\alpha(\omega - \omega_0) + g_\alpha(\omega + \omega_0)] +$$

$$+ \frac{\epsilon^2 \omega^2 a_0^4 x^2}{64 c_0^4} [4 g_\alpha(\omega) + \delta(\omega - 2\omega_0) + \delta(\omega + 2\omega_0) +$$

$$4 g_\alpha(\omega + 2\omega_0) + 4 g_\alpha(\omega - 2\omega_0)] . \qquad (5.32)$$

As distinct from the signal with phase fluctuations, signal detection is observed and low-frequency components appear (the first term in the second square brackets).

Problem 5.12

Find the simple wave spectrum using the expression for its Fourier transform (1.53) assuming that stationary noise is prescribed at the input with the characteristic function

$$\theta_2(\gamma_1, \gamma_2, \rho) = \langle e^{i\gamma_1 \Phi(\tau) + i\gamma_2 \Phi(\tau + \rho)} \rangle , \qquad (5.33)$$

where $\Phi(\tau) = u_0(x = 0, \tau)$. Consider the spectrum behaviour at the initial stage.

Solution For a stationary process the Fourier-transform $C(\omega)$ and the power spectrum $g(\omega)$ are related through

$$\langle C(\omega) C^*(\omega') \rangle = g(\omega) \delta(\omega - \omega') . \qquad (5.34)$$

Multiplying the simple wave Fourier-transform $C(x, \omega)$ of (5.33) by the complex conjugate $C^*(x, \omega')$ and averaging we have

$$\langle C(\omega)C^*(\omega') \rangle = (4\pi^2 \omega \omega' (\epsilon/c_0^2)^2 x^2)^{-1}$$

$$\int \int_{-\infty}^{\infty} [\theta_2(\omega(\epsilon/c_0^2)x), -\omega'(\epsilon/c_0^2)x; \xi_2 - \xi_1) - \theta_1(\omega(\epsilon/c_0^2)x) -$$

$$\theta_1(-\omega'(\epsilon/c_0^2)x) + 1] \exp(-i\omega\xi_1 + i\omega'\xi_2) \, d\xi_1 \, d\xi_2 \ . \tag{5.35}$$

Here $\theta_1(\gamma) = \langle \exp(i\gamma\Phi(\tau)) \rangle$ is a one-dimensional characteristic function. Proceeding to the integration with respect to $\xi = \xi_2 - \xi_1$ and ξ_1 and taking into account that

$$\frac{1}{2\pi} \int_{-\infty}^{+\infty} e^{i\omega\xi} \, d\xi = \delta(\omega) \ , \tag{5.36}$$

we obtain for the intensity spectrum

$$g(x, \omega) = (2\pi\omega^2 (\epsilon/c_0^2)^2 x^2)^{-1}$$

$$\int_{-\infty}^{+\infty} [\theta_2(\omega(\epsilon/c_0^2)x), -\omega(\epsilon/c_0^2)x; \xi) - \theta_1(\omega(\epsilon/c_0^2)x) -$$

$$\theta_1(-\omega(\epsilon/c_0^2)x) + 1] \exp(i\omega\xi) \, d\xi \ . \tag{5.37}$$

Let for simplicity $\langle \Phi \rangle = 0$, then expand the characteristic function into a series in terms of the moments:

$$\theta_2(\gamma_1, \gamma_2, \xi) = 1 - \frac{1}{2}\gamma_1^2\sigma^2 - \frac{1}{2}\gamma_2^2\sigma^2 - \gamma_1\gamma_2 B_0(\xi) + \dots \ .$$

$$\theta_1(\gamma) = 1 - \frac{1}{2}\gamma^2\sigma^2 + \dots \ ;$$

$$\sigma^2 = \langle \Phi^2 \rangle \ , \quad B_0(\xi) = \langle \Phi_0(\tau) \, \Phi_0(\tau + \xi) \rangle \ . \tag{5.38}$$

From (5.37), as $x \to 0$, we have

$$g(x, \omega) = \frac{1}{2\pi} \int_{-\infty}^{+\infty} B_0(\xi) e^{i\omega\xi} \, d\xi = g_0(\omega) \ , \tag{5.39}$$

where $g_0(\omega)$ is the signal spectrum at input.

Problem 5.13

Find the simple wave spectrum assuming that a stationary Gaussian noise with zero mean and the correlation function $B_0(\xi)$ is specified at the input.

Answer Making use of the expression for a characteristic function of the Gaussian process one can formulate the following expression out of (5.37)

$$g(x,\omega) = \frac{e^{-\omega^2(\epsilon/c_0^2)^2\sigma^2 x^2}}{2\pi\omega^2(\epsilon/c_0^2)^2 x^2} \int_{-\infty}^{+\infty} [e^{\omega^2(\epsilon/c_0^2)^2 x^2 B_0(\xi)} - 1]e^{i\omega\xi}\, d\xi \ , \quad (5.40)$$

where $\sigma^2 = B_0(0)$ is the input signal dispersion.
Remark: The two-point characteristic function of the Gaussian process can be readily written if we recall that the equality $\langle e^{i\alpha\gamma} \rangle = e^{i\gamma\langle\alpha\rangle - \gamma^2\sigma_\alpha^2/2}$ holds true for the Gaussian random value α; here σ_α^2 is dispersion: $\sigma_\alpha^2 = \langle(\alpha - \langle\alpha\rangle)^2\rangle$.

Problem 5.14

Assuming that the Gaussian signal correlation function is characterized by the single time scale $\tau_* = 1/\omega_0$ and is rendered in the form $B_0(\xi) = \sigma_0^2 \tilde{R}(\xi\omega_0)$, derive a dimensionless expression for the simple wave spectrum (5.40).

Answer

$$g(x,\omega) = \sigma_0^2\omega_0\tilde{g}(z,\Omega) \ , \quad \Omega = \omega/\omega_0 \ , \quad z = \frac{\epsilon}{c_0^2}\sigma_0\omega_0 x \ , \quad (5.41)$$

$$\tilde{g}(z,\Omega) = \frac{e^{-z^2\Omega^2}}{2\pi\Omega^2 z^2} \int_{-\infty}^{+\infty} [e^{z^2\Omega^2 \tilde{R}(\eta)} - 1]\, e^{i\Omega\eta}\, d\eta \ , \quad (5.42)$$

Problem 5.15

Analyze the evolution of the spectrum and the correlation function of a simple wave represented at the input by a quasimonochromatic signal with the correlation function

$$B_0(\xi) = \sigma^2 b_0(\xi) \cos(\omega_0 \xi) \ , \qquad b_0(\xi) = \tilde{b}(\Delta \xi) \ , \qquad (5.43)$$

where $b_0(\xi)$ is a slow (in the scale $\cos(\omega_0 \xi)$) function characterized by the scale $T = 1/\Delta_*$ such that $\mu = \Delta_*/\omega_0 \ll 1$.

Solution Using the change of variables (5.41) for a dimensionless spectrum we obtain from (5.42)

$$g(z, \Omega) = \frac{e^{-z^2 \Omega^2}}{2\pi \Omega^2 z^2} \int_{-\infty}^{+\infty} [e^{z^2 \Omega^2 \tilde{b}_0(\mu\eta) \cos \eta} - 1] \, e^{i\Omega\eta} \, d\eta \ . \qquad (5.44)$$

Making use of the exponent expansion in terms of modified Bessel functions $I_n(z)$ (3.25), one can write (5.44) as a sum of the spectra at the signal harmonics and the low-frequency component

$$g(z, \Omega) = g_0(z, \Omega) + \sum_{n=1}^{\infty} g_n(z, \Omega) \ , \qquad (5.45)$$

$$g_0(z, \Omega) = \frac{e^{-z^2 \Omega^2}}{2\pi \Omega^2 z^2} \int_{-\infty}^{+\infty} [I_0(z^2 \Omega^2 \tilde{b}_0(\mu\eta)) - 1] \, e^{i\Omega\eta} \, d\eta \ , \qquad (5.46)$$

$$g_n(z, \Omega) = \frac{2e^{-z^2 \Omega^2}}{2\pi \Omega^2 z^2} \int_{-\infty}^{+\infty} I_n(z^2 \Omega^2 \tilde{b}_0(\mu\eta)) \, e^{i\Omega\eta} \cos(n\eta) \, d\eta \ . \qquad (5.47)$$

Inasmuch as $\mu \ll 1$, the spectrum of the nth harmonic will be concentrated near $\Omega \approx n$ and in the expression (5.47) the arguments Ω can be replaced by n, leading to

$$g_n(z, \Omega) = \frac{2e^{-n^2 z^2}}{2\pi n^2 z^2} \int_{-\infty}^{+\infty} I_n(n^2 z^2 \tilde{b}_0(\mu\eta)) \, e^{i\Omega\eta} \cos(n\eta) \, d\eta \ . \qquad (5.48)$$

It immediately follows from (5.48) that the correlation function can be formulated as a sum of the individual harmonic correlation functions

$$B_n(z, \xi) = \frac{2e^{-n^2 z^2}}{n^2 z^2} I_n(n^2 z^2 \tilde{b}_0(\mu\eta)) \, \cos(n\xi) \ , \qquad (5.49)$$

and the low-frequency component. Using the Bessel function expansion one can demonstrate that the harmonic generation efficiency at the initial stage for noise is $n!$ times higher than for the regular signal (see Problem 5.5).

Problem 5.16

Employing the results of the previous problem, find an expression for the low-frequency spectrum portion emerging due to the modulated high-frequency signal detection and estimate the nth harmonic spectrum width $\Delta\omega_n$ at the initial stage assuming at the input

$$b_0(\xi) = \mathrm{e}^{-\frac{1}{2}\xi^2\Delta^2} \ , \qquad \tilde{b}_0 = \mathrm{e}^{-\frac{1}{2}\xi^2} \ . \tag{5.50}$$

Solution For the low-frequency component at $z < 1$ (where the simple wave approximation applies) $I_0(\gamma) = 1 + \gamma^2/4$ can be expanded leading to

$$g_0(z,\Omega) = \frac{1}{8\pi}z^2\Omega^2 \int_{-\infty}^{+\infty} \tilde{b}_0^2(\mu\eta)\,\mathrm{e}^{i\Omega\eta}\,d\eta = \frac{z^2\Omega^2}{4\sqrt{4\pi\mu^2}}\mathrm{e}^{-\Omega^2/(4\mu^2)} \ . \tag{5.51}$$

For the higher harmonics at $z \ll 1$ one can employ the expansion $I_n(z) \sim z^n$ and then

$$g_n(z,\Omega) \sim \int_{-\infty}^{+\infty} d\eta\, b_0^n(\mu\eta)\,\mathrm{e}^{i\Omega\eta}\cos(n\eta) \sim \mathrm{e}^{-(\Omega-n)^2/(2n\mu^2)} \ . \tag{5.52}$$

Thus $\Delta\Omega_n \sim \sqrt{n}\mu$ or $\Delta\omega_n \sim \Delta\sqrt{n}$. At $\Delta\omega_n \sim \omega_0$ the harmonic spectra merge.

Problem 5.17

Using the simple wave solution, demonstrate that for stationary noise the one-point probability distribution will be retained. Presume that the ergodicity conditions are satisfied.

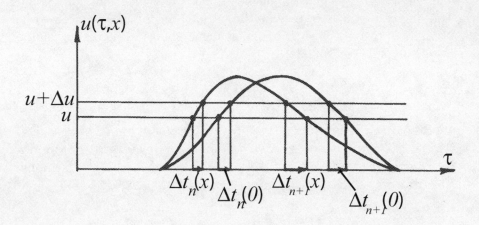

Figure 5.1: *Explanation of the conservation of one-point probability at strong profile distortion but at absence of shocks.*

Answer For the ergodic process the probability distribution coincides with the relative time of the process staying within the interval $[u, u + \Delta u]$ (Figure 5.1)

$$W(u, x) = \lim_{T \to \infty, \Delta u \to 0} (T \Delta u)^{-1} \sum \Delta t_n \ , \tag{5.53}$$

where T is the total interval length and Δt_n is the interval length where the function falls within the interval $[u, u + \Delta u]$. Due to nonlinear distortions each interval length will vary. For each profile point within the accompanying coordinate system the relationship $t = t_0 - (\epsilon/c_0^2)ux$ holds true

$$\Delta t_n(x) = \Delta t_n(0) \pm (\epsilon/c_0^2)\Delta u \cdot x \tag{5.54}$$

and hence, the sum of any two neighbouring time intervals will be constant: $\Delta t_n + \Delta t_{n+1} = constant$. Consequently the probability distribution will not change either. The probability distribution change results from the discontinuity formation.

Problem 5.18

Find the probability distribution of the harmonic input signal

$$u(\tau) = a\sin(\omega_0\tau + \phi) \tag{5.55}$$

with a random phase uniformly distributed within the interval $[-\pi, \pi]$. Consider the stage before discontinuity formation $(x < x_S = c_0^2/(\epsilon\omega_0 a))$, and that of the developed discontinuities $(x \gg x_S)$.

Answer For $x < x_S$

$$W(u, x) = \frac{1}{\pi\sqrt{a^2 - u^2}} \tag{5.56}$$

and for $x \gg x_S$

$$W(u, x) = \begin{cases} 1/(2\Delta V) & |u| < \Delta V \\ 0 & |u| > \Delta V \end{cases} \tag{5.57}$$

where $\Delta V = \pi c_0^2/(\epsilon\omega x)$.

Problem 5.19

Using the limiting solution of the Burgers equation in the case of infinitely small viscosity (Problem 3.10), show that stationary and continuous input noise is transformed into a sequence of saw-tooth pulses with the same slope at sufficiently large distances. Find the velocities of the discontinuities.

Solution Let the input noise have dispersion $\sigma_u^2 = \langle u_0^2(\tau) \rangle$ and be characterized by the scale τ_0. Then the characteristic curvature of the function $\beta S_0(\tau)$ entering the solution (3.37), (3.38) equals $\beta S_0''(\tau) \sim \beta\sigma_u/\tau_0$. The parabola α curvature in the same solution is $1/x$. For $\beta\sigma_u x/\tau_0 \gg 1$ parabola $\alpha(t, \tau, x)$ will be a smooth function of t in the scale $\beta S_0(t)$.

Because of this, the touching points between $\beta S_0(t)$ and $\alpha(t, \tau, x)$ will be close to some maxima $\beta S_0(t)$ (see Figure 5.2).

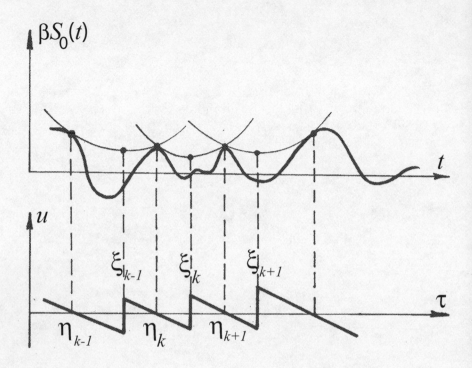

Figure 5.2: *Transformation of stationary noise into a sequence of sawtooth-shaped pulses having straight-line connections - all with the same inclination.*

The field $u(x, \tau)$ is completely defined by a system of critical parabolas, i.e. the parabolas having double touching points with function $\beta S_0(t)$. Therewith, the critical parabola centre coordinates define the position of discontinuities ξ_K and the critical parabola intersection points (coinciding with some maxima $\beta S_0(t)$) define the zeros η_K on the field $u(x, \tau)$. Indeed, within the interval between discontinuities ξ_K and ξ_{K+1}, the parabola α touches the function $\beta S_0(t)$ practically in the same point η_K which means that the field $u(x, \tau)$ has a universal structure in the intervals between discontinuities

$$u(x,\tau) = (\eta_K - \tau)/(\beta x) , \qquad \xi_K < \tau < \xi_{K+1} . \tag{5.58}$$

The discontinuity position is determined from the double touching condition between α and βS_0 and for the discontinuity coordinate one can write

$$\xi_K = \frac{\eta_K + \eta_{K-1}}{2} - \beta x \frac{S_0(\eta_K) - S_0(\eta_{K-1})}{\eta_K - \eta_{K-1}} . \tag{5.59}$$

Besides, the discontinuity travel speed is constant

$$V_K = \frac{d\xi_K}{dx} = -\beta \frac{S_0(\eta_K) - S_0(\eta_{K-1})}{\eta_K - \eta_{K-1}} . \tag{5.60}$$

Therefore, the field $u(x,\tau)$ profile at this stage appears as a set of inclined lines with the same slope $-1/\beta x$ emerging from the zeros $\tau = \eta_K$. These lines are connected by vertical lines - discontinuities having coordinates ξ_K. The distance between individual neighbouring discontinuities $\Delta_K = \xi_{K+1} - \xi_K$ can both increase or decrease. If Δ_K decreases the discontinuities merge to form a single one with an amplitude equal to the sum of the merged discontinuity amplitudes.

Problem 5.20

By supposing that a random field $u(x,\tau)$ is characterized by the only scale $\tau(x)$, estimate the growth of this scale due to discontinuity coalescence.

Solution In the case of random perturbations $u_0(\tau)$, the discontinuity velocities are random as well. Because of this, collision and sticking together of the discontinuities will occur which gives rise to the increase of the field $\tau(x)$ characteristic time scale. The estimate of the $\tau(x)$ growth can be obtained through writing an equation for the average frequency of the discontinuity (per unit time) $n(x)$ which is related to the external scale by means of $n(x) = 1/\tau(x)$. A decrease of $n(x)$ due to collisions is proportional both to the discontinuity number

$n(x)$ and to the ratio of the characteristic speed of the discontinuity approach $\Delta v = V_{K+1} - V_K$ to the characteristic distance between them

$$\frac{dn}{dx} = -n\frac{\Delta v}{\tau} = -n^2 \Delta v \ . \tag{5.61}$$

The discontinuity approach speed Δv can be assumed to be on the order of characteristic scattering of the discontinuity velocity $\langle \Delta v_K^2 \rangle$. Use of the expression for the discontinuity velocity (5.60) yields the following estimate

$$\langle \Delta v_K^2 \rangle \approx \langle v_K^2 \rangle \approx \beta^2 \frac{\langle (S_0(\eta + \tau) - S_0(\eta))^2 \rangle}{\tau^2} \ . \tag{5.62}$$

Or, with $B_0(\tau) = \langle u_0(\eta + \tau) u_0(\eta) \rangle$ and $B_0(0) = \sigma_0^2$ being the specified input signal correlation function, then

$$\langle \Delta v^2 \rangle = \beta^2 n \int_0^{1/n} (1 - \eta n) B_0(\eta)\, d\eta = \begin{cases} n\sigma_0^2 \tau_0 \ , & D \neq 0 \ , \\ n^2 \sigma_0^2 \tau_0^2 \ , & D = 0 \ . \end{cases} \tag{5.63}$$

Here $D = \int_0^\infty B_0(\eta)\, d\eta$ is the value of the initial perturbation spectrum at zero frequency. (For $D \neq 0$ the reference correlation time τ_0 is determined from the condition $D = \sigma_0^2 \tau_0$ and for $D = 0$ from the condition $\int_0^\infty \eta B_0(\eta)\, d\eta = -\sigma_0^2 \tau_0^2$). Substituting (5.63) into (5.61) yields the following estimates for the external scale growth

$$\tau(x) \approx \begin{cases} \tau_0 (x/x_S)^{2/3} \ , & D \neq 0 \ , \\ \tau_0 (x/x_S)^{1/2} \ , & D = 0 \ , \end{cases} \tag{5.64}$$

where $x_S = \tau_0/\beta\sigma_0$ is the characteristic length of the nonlinear effect.

Problem 5.21

Assuming that statistical properties of the intense noise tend to be self-similar, write an expression for the wave power spectrum (a) and, by using (5.64), evaluate the field energy at the stage of developed discontinuities (b).

Answer a) $g(x,\omega) = \frac{\tau^3(x)}{\beta^2 x^2}\tilde{g}(\omega\tau(x))$, where $\tilde{g}(\Omega)$ is a universal dimensionless function.

b) $\qquad \langle u^2(x,\tau)\rangle \approx \frac{\tau^2(x)}{\beta^2 x^2} \approx \sigma_0^2 \begin{cases} (x_S/x)^{2/3} & D \neq 0 \\ (x_S/x) & D = 0 \end{cases}$ (5.65)

Therefore, because of the discontinuity coalescence, the noise energy falls off slower than for a harmonic input signal for which $\langle u^2\rangle \sim x^{-2}$. Remark: Elaborate descriptions of the problems concerning intense noise wave evolution are offered in monographs and reviews [4,13-15].

Problem 5.22

Using the expression for the simple wave Fourier transform (1.53) and the expansion [22]

$$\exp(iz\cos\phi) = \sum_{k=-\infty}^{+\infty} i^k J_k(z)e^{ik\phi} , \qquad (5.66)$$

where $J_k(z)$ is the Bessel function, find the intensity spectrum of an acoustic wave appearing with an input mix of noise $\eta(t)$ and signal $\Psi(t) = A\cos(\omega_0 t + \phi)$ where the phase ϕ is uniformly distributed over the interval $[-\pi, \pi]$. Assume that $\eta(t)$ has Gaussian distribution with correlation function $B_\eta(\tau)$. Investigate the damping of discrete components due to nonlinear interaction with noise. Analyze the spectrum form provided the noise is of low frequency compared with the signal.

Answer The power spectrum can be readily formulated as

$$g(x,\omega) = \sum_{k=-\infty, k\neq 0}^{+\infty} \frac{J_k(k\beta\omega_0 Ax)}{k^2\beta^2\omega_0^2 x^2} e^{-k^2\beta^2\omega_0^2\sigma_0^2 x^2} \delta(\omega - k\omega_0) +$$

$$+\frac{J_0^2(\beta\omega Ax)}{2\pi\beta^2\omega^2 x^2} e^{-\beta^2\omega^2\sigma_0^2 x^2} \int_{-\infty}^{+\infty} [e^{\beta^2\omega^2 x^2 B_0(\xi)} - 1]e^{i\omega\xi} d\xi +$$

$$+ \sum_{k=-\infty, k\neq 0}^{+\infty} \frac{J_k^2(\beta\omega Ax)}{2\pi\beta^2\omega^2 x^2} e^{-\beta^2\omega^2\sigma_0^2 x^2} \int_{-\infty}^{+\infty} [e^{\beta^2\omega^2 x^2 B_0(\xi)} - 1]e^{i(\omega - k\omega_0)\xi} d\xi ,$$

$$(5.67)$$

where $B_0(\xi) = \langle\eta(\tau)\eta(\tau+\xi)\rangle$, $\sigma_0^2 = B_0(0)$.

Here the first sum describes harmonic amplitudes which are attenuated due to interaction with noise such that the damping decrement is defined by the total noise power σ_0^2 and increases with harmonic number.

The second term stands for the noise spectrum which is distorted because of the interaction with regular signals. And finally, the last sum represents new spectrum components emerging due to signal-noise interaction. If the characteristic noise frequency γ is much lower than the signal frequency ω_0, then newly emerging components will be located near the regular signal harmonics. For the components appearing near the kth harmonic one can write from (5.67)

$$g_k(x,\omega) \approx \frac{A_k^2}{2\pi} e^{-k^2\beta^2\omega_0^2\sigma_0^2 x^2} \int_{-\infty}^{+\infty} [e^{k^2\beta^2\omega_0^2 x^2 B_0(\xi)} - 1] e^{i(\omega - k\omega_0)\xi} d\xi \ , \quad (5.68)$$

where $A_k = J_k(k\beta\omega_0 Ax)/(k\beta\omega_0 x)$ is the amplitude of the kth harmonic. The spectrum shape is defined by the value $km = k\beta\omega_0\sigma_0 x$.

Problem 5.23

Demonstrate that if the distortions of the low frequency noise $\eta(\tau)$ can be disregarded, then for the newly emerged component spectrum the expression (5.67) will hold true at the discontinuous stage as well, with A_k being also the amplitudes of the discontinuous wave harmonics.

Solution If $\eta(\tau)$ has a constant value, then the solution of the Burgers equation will be described by (3.14). The same formula will be approximately valid provided $\eta(\tau)$ is a slow function and then

$$u(x,\tau) = \eta(\tau) + \Pi(\tau + \beta\eta(\tau)x, x) \ , \quad (5.69)$$

where $\Pi(x,\tau)$ is the high-frequency wave field. Thus, nonlinear interaction results in a modulation of the high-frequency wave harmonics. Assuming $\eta(\tau)$ to be a Gaussian signal one can obtain an expression for the spectrum (5.68).

Therefore, the value entering the previous problem answer defines the phase fluctuations. For $mk \ll 1$ the spectrum (5.67) implies that the form of the newly appearing harmonics repeats the low-frequency noise

$$g_k(x, \omega) \approx A_k \cdot k^2 \beta^2 \omega_0^2 x^2 g_\eta(\omega) \ , \tag{5.70}$$

$$g_\eta \omega) = \frac{1}{2\pi} \int_{-\infty}^{+\infty} B_0(\xi) e^{i\omega\xi} \, d\xi \ , \tag{5.71}$$

and for $mk \gg 1$ the spectrum has a universal form.

Indeed, expanding $B_0(\tau) = 1 - \gamma^2 \tau^2 / 2 + \ldots$ and making use of the saddle-point method we obtain from (5.68)

$$g_k(x, \omega) \approx \frac{A_k^2}{\sqrt{2\pi}\gamma k\beta\omega_0\sigma x} \exp[-\frac{(\omega - k\omega_0)^2}{2(\gamma k\beta\omega_0\sigma x)^2}] \ . \tag{5.72}$$

Consequently, this spectrum shape is Gaussian, and its width $\gamma mk \gg \gamma$ is much greater than the width of the low-frequency noise spectrum.

Chapter 6

NONLINEAR NONDESTRUCTIVE TESTING

Problem 6.1

A plane wave $p_+(t - x/c)$ propagates in a linear medium along the x-axis and reflects from a layer located in the domain $0 < x < h$ near a rigid immovable boundary (see Figure 6.1). The thickness h is small in comparison with the wave length. The deformation of the layer caused by the acoustic pressure $p_{ac}(t)$ obeys the equation

$$\hat{L}(X) = p_{ac}(t) \ , \tag{6.1}$$

where $X(T)$ is the displacement of the left boundary of the layer from its equilibrium position $x = 0$, and $\hat{L}$ is a linear or nonlinear operator. Derive:

1. The equation of motion (6.1) containing, instead of $p_{ac}(t)$, the known incident wave $p_+(t)$.

2. The equation describing the wave $p_-(t + x/c)$ reflected from the layer.

Solution The acoustic pressure field in the domain $-\infty < x < 0$ consists of two counter-propagating waves:

$$p_{ac}(t) = p_+(t - x/c) + p_-(t + x/c) \ , \tag{6.2}$$

where $p_+(t)$ is the temporal shape of the incident wave, and $p_-(t)$ is the shape of the reflected wave. Using the linear acoustic equation following from the Euler equation of motion of fluid

$$\rho c \frac{\partial u_{ac}}{\partial t} = -\frac{\partial p_{ac}}{\partial x} \ , \tag{6.3}$$

one can determine the field of vibration velocity corresponding to the pressure (6.1):

$$\rho c u_{ac} = p_+(t - x/c) - p_-(t + x/c) \ . \tag{6.4}$$

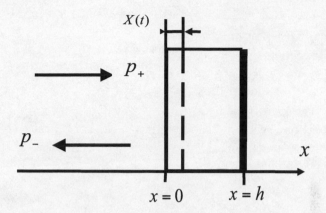

Figure 6.1: *Reflection of a plane wave from a layer located next to a rigid immovable boundary.*

Because the velocity of the left boundary dX/dt must be equal to the particle velocity of the medium at this boundary,

$$u_{ac}(x = 0, t) = dX/dt \ , \tag{6.5}$$

equations (6.2) and (6.4) are reduced to the following system:

$$p_{ac}(t) = p_+(t) + p_-(t) \ , \qquad \rho c \, dX/dt = p_+(t) - p_-(t) \ . \tag{6.6}$$

Now we can express the acoustic pressure through the incident wave:

$$p_{ac} = 2p_+(t) - \rho c \, dX/dt \ , \tag{6.7}$$

and derive the equation of motion:

$$\hat{L}(X) + \rho c \frac{dX}{dt} = 2p_+(t) \ . \tag{6.8}$$

Because the incident wave $p_+(t)$ is known, one can calculate the displacement $X(t)$ from the inhomogeneous equation (6.8) and determine the form of the reflected wave (see (6.6)):

$$p_-(t) = p_+(t) - \rho c \, dX/dt \ . \tag{6.9}$$

The next problems will make the general results (6.8) and (6.9) more specific.

Problem 6.2

Using the results of Problem 6.1, calculate the amplitude of the second harmonic wave which appears at the reflection of an incident harmonic wave from a weakly nonlinear layer. The deformation of the layer obeys the law

$$\hat{L}(X) = E\frac{X}{h}(1 + \epsilon\frac{X}{h}) \ , \tag{6.10}$$

where E is the Young modulus, and ϵ is the nonlinearity.

Solution Equation (6.8) with account for (6.10) takes the form

$$\rho c\frac{dX}{dt} + E\frac{X}{h}(1 + \epsilon\frac{X}{h}) = 2P_1\cos(\omega t) \ . \tag{6.11}$$

Here P_1 is the amplitude of incident harmonic wave having frequency ω. Because nonlinearity is weak equation (6.11) can be solved by the successive approximations method. The first approximation

$$\rho c\frac{dX^{(1)}}{dt} + \frac{E}{h}X^{(1)} = P_1(e^{i\omega t} + e^{-i\omega t}) \ , \tag{6.12}$$

corresponds to the linear problem ($\epsilon = 0$). The solution to equation (6.12) describes vibration at frequency ω of the incident wave:

$$X^{(1)} = P_1\frac{h}{E}\frac{1}{1 + iG}e^{i\omega t} + c.c. \ , \qquad G = \frac{\omega h}{c}\frac{\rho c^2}{E} \ , \tag{6.13}$$

where "c.c." means a "complex conjugated" term with respect to the directly predecessing term. The second harmonic is governed by the equation of second approximation:

$$\rho c\frac{dX^{(2)}}{dt} + \frac{E}{h}X^{(2)} = -\epsilon E\frac{X^{(1)2}}{h^2} \ . \tag{6.14}$$

Using solution (6.13) of the first approximation, we calculate the right-hand-side of equation (6.14):

$$-\epsilon E \frac{X^{(1)2}}{h^2} = -2\epsilon \frac{P_1^2}{E} \frac{1}{1+G^2} - \epsilon \frac{P_1^2}{E} \Big[\frac{1}{(1+iG)^2} e^{i2\omega t} + c.c. \Big] . \quad (6.15)$$

One can see that the external force deforming the layer contains a constant term producing static deformation, and a term oscillating at the second harmonic frequency. The static displacement is calculated by substitution of the static term in (6.15) into equation(6.14):

$$\frac{X_{stat}^{(2)}}{h} = -2\epsilon \frac{P_1^2}{E^2} \frac{1}{1+G^2} . \quad (6.16)$$

For the second harmonic equation (6.14) takes the form

$$\rho c \frac{dX^{(2)}}{dt} + \frac{E}{h} X^{(2)} = -\epsilon \frac{P_1^2}{E} \Big[\frac{1}{(1+iG)^2} e^{i2\omega t} + c.c. \Big] . \quad (6.17)$$

The solution to equation (6.17) is

$$\frac{X_{2\omega}^{(2)}}{h} = -\epsilon \Big(\frac{P_1}{E}\Big)^2 \frac{1}{(1+iG)^2(1+i2G)} e^{i2\omega t} + c.c. . \quad (6.18)$$

Using formula (6.9) we calculate the second harmonic in the reflected pressure wave

$$p_-(t) = -\rho c \frac{dX_{2\omega}^{(2)}}{dt} = -i\epsilon \Big(\frac{P_1}{E}\Big)^2 \frac{2\omega \rho c h}{(1+iG)^2(1+i2G)} e^{i2\omega t} + c.c. . \quad (6.19)$$

The amplitude of this wave equals

$$P_2 = \Big| \rho c \frac{dX_{2\omega}^{(2)}}{dt} \Big| = 4\epsilon \frac{P_1^2}{E} \frac{G}{(1+G^2)\sqrt{1+4G^2}} . \quad (6.20)$$

The ratio of dimensionless acoustic Mach number (reminder: $M = P/(\rho c^2)$) of the second harmonic to the squared Mach number of the first harmonic equals to

$$\frac{M_{2\omega}}{M_\omega^2} = 4\epsilon (kh)\Big(\frac{\rho c^2}{E}\Big)^2 \Big\{ \Big[1 + \Big((kh)\frac{\rho c^2}{E}\Big)^2 \Big] \sqrt{1 + 4\Big[(kh)\frac{\rho c^2}{E}\Big]^2} \Big\}^{-1} . \quad (6.21)$$

Here $k \equiv \omega/c = 2\pi/\lambda$ is the wave number. For a thin layer, $kh \ll 1$, the {}-bracket in equation(6.21) equals to unity, and

$$\frac{M_{2\omega}}{M_\omega^2} = 4\epsilon(kh)\left(\frac{\rho c^2}{E}\right)^2 . \qquad (6.22)$$

It follows from the result (6.22), that at a given amplitude of incident wave the reflected second harmonic increases with increase in three quantities:

1) Nonlinearity ϵ of the layer.
2) Width of layer kh, measured in wave length units.
3) Ratio of compressibilities $\rho c^2/E$ between nonlinear and linear media. If the nonlinear medium is much more compressible and $\rho c^2/E \gg 1$, the amplitude of the second harmonic increases many times. This phenomenon offers the possibility to perform nondestructive testing to detect voids (or cracks) in solids and gas bubbles in liquids; these internal defects are much softer than the surrounding medium.

Problem 6.3

Calculate the total temporal and spectral response of the nonlinear layer described in Problem 6.1 for a harmonic incident wave. The deformation obeys the law

$$\hat{L}(X) = \frac{E}{2\epsilon}\left(\exp(2\epsilon\frac{X}{h}) - 1\right) . \qquad (6.23)$$

At small deformations, $\epsilon X/h \ll 1$, the law (6.23) transforms to a quadratic nonlinear dependence (see (6.10)). But now, the deformation is large and the successive approximation method is not applicable.

Solution Equation (6.8) of the Problem 6.1 with account for the concrete law (6.23) takes the form

$$\rho c\frac{dX}{dt} + \frac{E}{2\epsilon}\left(\exp(2\epsilon\frac{X}{h}) - 1\right) = 2P_1\sin(\omega t) . \qquad (6.24)$$

Equation (6.24) has a remarkable property. It can be linearized by the substitution

$$X = -\frac{h}{2\epsilon} \ln Y \ .$$ (6.25)

Using (6.25), we reduce the nonlinear (6.24) to a linear equation for the new variable Y:

$$\frac{dY}{d\tau} + \frac{1}{G}(1 + b\sin\tau)Y = \frac{1}{G} \ .$$ (6.26)

For simplicity, the following notations are used:

$$G = \frac{wh}{c}\frac{\rho c^2}{E} \ , \qquad b = 4\epsilon\frac{P_1}{E} \ , \qquad \tau = \omega t \ .$$ (6.27)

Seeking for the solution to the linear equation(6.26) by variation of the constant

$$Y = C(\tau)\exp(-\frac{\tau}{G} + \frac{b}{G}\cos\tau) \ ,$$ (6.28)

we obtain the integral

$$C(\tau) = \frac{1}{G}\int_{-\infty}^{\tau} \exp(\frac{\tau_1}{G} - \frac{b}{G}\cos\tau_1)\, d\tau_1 \ .$$ (6.29)

This integral can be calculated by means of expansion (3.25) [22]

$$\exp(-\frac{b}{G}\cos\tau_1) = I_0(\frac{b}{G}) + 2\sum_{n=1}^{\infty}(-1)^n I_n(\frac{b}{G})\cos(n\tau_1) \ .$$ (6.30)

Here, in formula (6.30), I_n are modified Bessel functions. Using expansion (6.30), we reduce the calculation (6.29) to simple integrals

$$\int_{-\infty}^{\tau} e^{\tau_1/G}\cos(n\tau_1)\, d\tau_1 = \frac{Ge^{T/G}}{\sqrt{1 + n^2 G^2}}\cos[n\tau - \arctan(nG)] \ .$$ (6.31)

Consequently,

$$C = e^{T/G}\{I_0(\frac{b}{G}) + 2\sum_{n=1}^{\infty}(-1)^n I_n(\frac{b}{G})\frac{\cos[n\tau - \arctan(nG)]}{\sqrt{1 + n^2 G^2}}\} \ ,$$ (6.32)

and

$$-\frac{2\epsilon}{h}X = \frac{b}{G}\cos\tau + \ln\{I_0(\frac{b}{G}) + 2\sum_{n=1}^{\infty}(-1)^n I_n(\frac{b}{G})\frac{\cos[n\tau - \arctan(nG)]}{\sqrt{1+n^2G^2}}\}$$

(6.33)

If nonlinearity tends to zero, $\epsilon \to 0$, $b \to 0$, this nonlinear solution

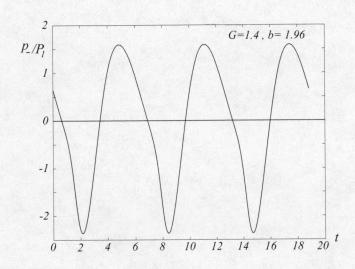

Figure 6.2: *Periodic strongly nonlinear temporal response of a thin layer to a harmonic incident signal.*

(6.33) tends to the linear response

$$X = P_1\frac{2h}{E}\frac{\sin(\omega t - \arctan G)}{\sqrt{1+G^2}}$$

(6.34)

in a form similar to formula (6.13) in Problem 6.2. Now, using formula (6.9) of Problem 6.1, we calculate the shape of the reflected wave which is the total temporal response

$$\frac{p_-}{P_1} = -\sin(\tau) - 4\frac{G}{b}\frac{\sum_{n=1}^{\infty}n(-1)^n I_n(\frac{b}{G})\frac{\sin[n\tau - \arctan(nG)]}{\sqrt{1+n^2G^2}}}{I_0(\frac{b}{G}) + 2\sum_{n=1}^{\infty}(-1)^n I_n(\frac{b}{G})\frac{\cos[n\tau - \arctan(nG)]}{\sqrt{1+n^2G^2}}}$$

(6.35)

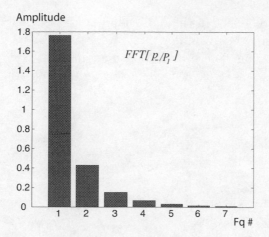

Figure 6.3: *The spectrum of the response shown in Figure 6.2.*

The total response is shown in Figure 6.2 for parameters $G = 1$ and $b = 3$. The corresponding spectrum is shown in Figure 6.3[3].

Problem 6.4

Some methods of nondestructive testing use a probing pulse signal with a wide-band frequency spectrum. Such a pulse can be excited simply by a hammer or by fast thermal expansion of a layer absorbing an incident laser radiation. To evaluate the material nonlinearity, it is sometimes convenient to use a combined probing signal, like a harmonic high-frequency sound wave and an impact pulse. Calculate the vibration spectrum of the boundary of the layer shown in Figure 6.1 produced by a pulse-harmonic wave interaction.

Solution Using the successive approximation method to calculate the result of interaction of pulse and harmonic signal, rewrite the equation (6.14) of Problem 6.2 as

$$\rho c \frac{dX^{(2)}}{dt} + \frac{E}{h} X^{(2)} = -\frac{\epsilon E}{h^2} (X^{(P)} + X^{(H)})^2 \rightarrow -\frac{2\epsilon E}{h^2} X^{(P)} X^{(H)} \ . \quad (6.36)$$

---

[3]Similar responses were calculated in the paper: C.M Hedberg, and O.V. Rudenko, J.Acoust.Soc.Am. **110**(5), 2340-2350, 2001

Only one term is kept in the right-hand-side of equation (6.36) corresponding to the chosen interaction. Here $X^{(P)}$, $X^{(H)}$, $X^{(2)}$ are three types of vibration: pulse, harmonic and nonlinear, correspondingly. Their spectra $S^{(P)}$, $S^{(H)}$, $S^{(2)}$ are given by the Fourier transform:

$$X^{(P)} = \int_{-\infty}^{\infty} S^{(P)}(\omega_1)e^{i\omega_1 t}\,d\omega_1 \;, \quad X^{(H)} = \int_{-\infty}^{\infty} S^{(H)}(\omega_2)e^{i\omega_2 t}\,d\omega_2 \;,$$

$$X^{(2)} = \int_{-\infty}^{\infty} S^{(2)}(\omega_3)e^{i\omega_3 t}\,d\omega_3 \qquad . \tag{6.37}$$

After substitution of spectral expansions (6.37) into equation(6.36) we have an integral relation

$$\int_{-\infty}^{\infty} (i\omega_3\rho c + \frac{E}{H})S^{(2)}e^{i\omega_3 t}\,d\omega_3 =$$

$$-\frac{2\epsilon E}{h^2} \int_{-\infty}^{\infty}\int_{-\infty}^{\infty} S^{(P)}S^{(H)}\,e^{i\omega_1 t + i\omega_2 t}d\omega_1 d\omega_2 \;. \tag{6.38}$$

Multiplying (6.38) by $\exp(-i\omega t)$ and integrating over time from minus to plus infinity, we derive the following relation between frequency spectra:

$$S^{(2)}(\omega) = -\frac{2\epsilon}{h}(1+i\frac{\omega h}{c}\frac{\rho c^2}{E})^{-1} \int_{-\infty}^{\infty} S^{(P)}(\omega_1)S^{(H)}(\omega-\omega_1)\,d\omega_1 \;. \tag{6.39}$$

At the derivation of (6.39) the well-known properties of the delta-function are used:

$$\int_{-\infty}^{\infty} \exp[i(\omega_3-\omega)t]\,dt = 2\pi\cdot\delta(\omega_3-\omega) \;, \quad \int_{-\infty}^{\infty} S(\omega_3)\delta(\omega_3-\omega)\,d\omega = S(\omega) \;. \tag{6.40}$$

One can see that the result (6.39) of interaction between two vibrations is proportional to the convolution of their spectra

$$S^{(2)}(\omega) \sim \int_{-\infty}^{\infty} S^{(P)}(\omega_1)S^{(H)}(\omega - \omega_1)\,d\omega_1 = S^{(P)}(\omega) \otimes S^{(H)}(\omega) \;. \tag{6.41}$$

For a harmonic signal

$$X^{(H)}(t) = X_1 \cos(\omega_0 t)$$

$$S^{(H)} = \frac{X_1}{2}\delta(\omega_0 - \omega) + \frac{X_1}{2}\delta(\omega_0 + \omega) \tag{6.42}$$

the spectrum of nonlinear interaction is

$$S^{(2)}(\omega) = -\frac{\epsilon}{h}\frac{X_1}{\left(1 + i\frac{\omega h}{c}\frac{\rho c^2}{E}\right)}[S^{(P)}(\omega - \omega_0) + S^{(P)}(\omega + \omega_0)] . \tag{6.43}$$

According to solution (6.43), the spectrum $S^{(2)}$ consists of two side-bands located near the spectral line $\omega = \omega_0$ of the high-frequency signal. Each sideband has the shape of the pulse spectrum $S^{(P)}(\omega)$. One, $S^{(P)}(\omega - \omega_0)$ lies to the left of spectral line, and another, $S^{(P)}(\omega + \omega_0)$ lies to the right of it. The initial spectrum is shown in Figure 6.4a, and the resulting spectrum of interaction is shown in Figure 6.4b.

Problem 6.5

At nondestructive testing of a layer, shown in Figure 6.1, a harmonic probing signal is used. Indicate parameters of this layer which can be measured during linear and nonlinear tests, based on the results of Problem 6.2.

Solution The solution to the first approximation (6.13), Problem 6.2, can be written as a real function (see also (6.34), Problem 6.3):

$$X^{(1)} = 2P_1\frac{h}{E}\frac{\cos(\omega t - \arctan G)}{\sqrt{1 + G^2}} , \qquad G = \frac{\omega h}{c}\frac{\rho c^2}{E} . \tag{6.44}$$

Using (6.1) and formula (6.9), Problem 6.1, we calculate the reflected pressure wave

$$p_-^{(1)} = p_+ - \rho c\frac{\partial X^{(1)}}{\partial t} = P_1[\cos(\omega t) + \frac{2G}{\sqrt{1 + G^2}}\sin(\omega t - \arctan G)] . \tag{6.45}$$

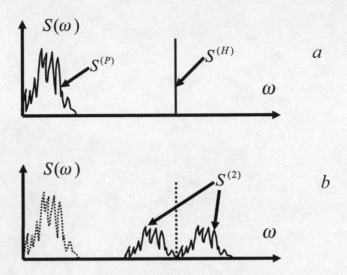

Figure 6.4: *Spectrum of the incident (probing) signal consisting of a high-frequency harmonic wave and pulse (a); and the resulting spectrum (b) formed as a result of interaction between the narrow-band and wide-band waves.*

After simple trigonometric transformations the reflected linear wave (6.45) can be rewritten in a simple form

$$p_-^{(1)} = P_1 \cos(\omega t - \arctan \frac{2G}{1 - G^2}) \; . \qquad (6.46)$$

One can see that the reflection coefficient always equals unity. This fact follows from the assumption that there is no energy loss in the layer. So, during linear tests only the phase shift $\arctan[2G/(1 - G^2)]$ needs to be measured and, consequently, one may obtain the parameter combination

$$G = \frac{\omega h}{c} \frac{\rho c^2}{E} \; . \qquad (6.47)$$

Of course, the parameters ρ, c of the linear medium in which the probing wave propagates are known. The frequency ω is predetermined as well. Consequently, the measured magnitude of G (6.47) offers the possibility to evaluate Young's module E, if the thickness h of layer is known, or, in the opposite case, acoustically measure h if the module E is known in advance. If both quantities are unknown, only their ratio $h/E = G/(\omega\rho c)$ can be evaluated.

During a nonlinear test the amplitude of the second harmonic (6.21), Problem 6.2, is measured. Because both the combination G and the ratio h/E are determined after a linear test, we can evaluate the coefficient of nonlinearity ϵ. At this evaluation the amplitude of the incident wave P_1 (or corresponding Mach number M_ω) must be known as well.

Problem 6.6

A layer of air bubbles forms at the boundary of the water tank filled with fresh water. It is known that the sound velocity of a bubbly liquid c_{eff} falls off with increase in volume concentration ν of small bubbles having resonant frequency much higher than the frequency of the acoustic wave. The effective sound velocity is given by the formula

$$c_{\text{eff}} = \frac{c}{1 + \nu\beta} \ , \qquad \beta = \frac{c^2\rho}{c_a^2\rho_a} \ . \tag{6.48}$$

In (6.48) c, ρ are the sound velocity and the density of water, and c_a, ρ_a are sound velocity and density of air. Determine the volume concentration of bubbles and the thickness of the layer using reflected waves at first and second harmonics. The incident wave is harmonic.

Solution First of all, we evaluate the ratio (6.48) of compressibility of air to the compressibility of water. In tables of physical quantities the sound velocities and densities are given: $\rho = 1000$ kg/m^3, $\rho_a = 1.3$ kg/m^3, $c = 1500$ m/s, $c_a = 330$ m/s. For these data the ratio (6.48) equals to $\beta \approx 1.6 \cdot 10^4$. One can see that the effective sound velocity c_{eff} can decrease 17 times at small volume concentration $\nu = 10^{-3}$ of bubbles. At increase in the gas concentration over $\nu = 10^{-3}$ the velocity c_{eff} can decrease much more, down to its minimum value of $c_{\text{eff}} = 23.8$ m/s at $\nu = 0.5$. This value is much less than the sound velocity in both water and air. Such highly gas-saturated water is out of our consideration here; formula (6.48) is valid only at significantly smaller concentration $\nu \sim 10^{-2} - 10^{-5}$. Because gas concentration is small, the density of air-water mixture is about the same as the density of water, and instead of the Young's modulus in formula (6.47), Problem 6.5, we can put $E = c_{\text{eff}}^2 \rho$,

$$G = \frac{\omega h}{c}(1 + \nu\beta)^2 \ . \tag{6.49}$$

As was shown in Problem 6.5, the parameter G can be be measured in a linear test. However, two unknown quantities are present in G, namely, the thickness of the layer h and the air concentration ν, which cannot be evaluated from equation (6.49). Performing the nonlinear measurement, we determine (see (6.22), Problem 6.2)

$$R \equiv \frac{M_{2\omega}}{M_\omega^2} = 4\epsilon(\frac{\omega}{c}h)(1 + \nu\beta)^4 \ . \tag{6.50}$$

We remind that $\epsilon = 1.2$ is the nonlinearity of air. From the system of two equations (6.49) and (6.50) one can calculate the unknown thickness and the air concentration

$$\nu = \frac{1}{\beta}(\sqrt{\frac{R}{4\epsilon G}} - 1) \ , \qquad h = \frac{4\epsilon G^2}{R}\frac{c}{\omega} \ , \tag{6.51}$$

which are expressed through known and measured quantities.

Problem 6.7

Estimate the thickness of layer and the volume concentration of air in Problem 6.6 for the experimentally measured Mach numbers $M_{2\omega} = 10^{-8}$, $M_\omega = 10^{-4}$ with the parameter $G = 10^{-1}$. The frequency of the probing wave was 10 kHz.

Answer The thickness of the layer is about $h \sim 1$ mm, and volume concentration of air is about $\nu \sim 10^{-5}$.

Problem 6.8[4]

To evaluate the tensile strength and the compression strength of a solid using an ultrasonic test, both the quadratic (A_1) and the cubic (A_2) nonlinear modules must be taken into account in the stress-strain relationship

$$\sigma(e) = E(e - A_1 e^2 - A_1 A_2 e^3) \ . \tag{6.52}$$

Here σ is the stress, e is the strain, and E is the linear elastic module. The dependence (6.52) is shown in Figure 6.5, where nonlinearity is described by positive constants $(A_1 > 0$ and $A_2 > 0)$ as is typical for solids. Derive the tensile and compression strengths expressed in the linear and nonlinear modules.

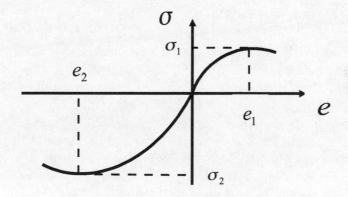

Figure 6.5: *The stress-strain relationship used in the evaluation of limiting breaking strain (6.53).*

---

[4]This problem is based on ideas of the paper: L.K.Zarembo, V.A.Krasilnikov, i.e. Shkolnik. Defectoscopiya (Russian J.of Nondestructive Testing) **10**, 1989

Solution The breaking strains are defined by the equation (see Figure 6.5):

$$\frac{d\sigma}{de} = 0 \ , \qquad 2A_1A_2\,e^2 + 3A_1\,e - 1 = 0 \ . \tag{6.53}$$

It follows from equation (6.53) that tensile and compression strains are equal to

$$e_1 = \frac{-A_1 + \sqrt{A_1^2 + 3A_1A_2}}{3A_1A_2} > 0 \ , \qquad e_2 = \frac{-A_1 - \sqrt{A_1^2 + 3A_1A_2}}{3A_1A_2} < 0 \ . \tag{6.54}$$

Substituting (6.54) into (6.52), one can calculate the tensile σ_1 and compression σ_2 strengths. If the cubic nonlinearity is weak in comparison with the quadratic one, $A_2 \ll A_1$, these strengths are given by the simple formulas

$$\sigma_1 \approx \frac{E}{2A_1} \ , \qquad \sigma_2 \approx -\frac{2E}{3A_2}(1 + \frac{2}{9}\frac{A_1}{A_2}) \ . \tag{6.55}$$

One can see, that the compression strength absolute value is much greater than the tensile strength: $|\sigma_2| \gg \sigma_1$. This inequality is true for most solids. It means that much less force is needed to break a fragile body by tension (or plastically deform it if the body is ductile) than by compression. One can see from the formulas (6.54) and (6.55), that the cubic nonlinearity is responsible for the appearance of the second root of equation (6.53), e_2, σ_2. Consequently, the tensile strength can be evaluated if the quadratic nonlinearity is known, but evaluation of compression strength needs numerical data on the cubic nonlinearity. It is possible to evaluate the quadratic nonlinearity using measurement of the amplitude of the second harmonic wave (see Problems 6.2 and 6.5). To evaluate the cubic nonlinearity, the nonlinear shift of the natural frequency of a resonator can be used (see the next Problem 6.9).

Problem 6.9

A solid has a form of parallelepiped. One of its edges of length L is parallel to the x-axis of a Cartesian coordinate system. At the side $x = 0$ of the sample an acoustic source is placed which vibrates harmonically in time, and the boundary $x = L$ is free. The vibration (in strain) is governed by the wave equation

$$\frac{\partial^2 e}{\partial t^2} = c^2 \frac{\partial^2}{\partial x^2} \left[\frac{\sigma(e)}{E} \right] \, , \quad c = \sqrt{\frac{E}{\rho}} \, . \tag{6.56}$$

By using the nonlinear stress-strain relationship (see (6.52)), calculate the nonlinear frequency shift of the fundamental mode.

Solution The wave equation (6.56) with account for nonlinear dependence $\sigma(e)$ has the form

$$\frac{\partial^2 e}{\partial t^2} - c^2 \frac{\partial^2 e}{\partial x^2} = -c^2 \frac{\partial^2}{\partial x^2} (A_1 e^2 + A_1 A_2 e^3) \, . \tag{6.57}$$

A linear vibration corresponds to the right-hand-side of equation (6.57) being zero. The linear solution

$$e = e_0 \sin(\omega_0 t) \cos(k_0 x) \tag{6.58}$$

is written here in form (6.58) with account for a vibrating boundary at $x = 0$; it vibrates according to the law $e(x = 0, t) = e_0 \sin(\omega t)$. To satisfy the zero boundary condition at the other boundary $x = L$, we put $\cos(k_0 L) = 0$. This condition determines the set of natural frequencies $k_0 L = \frac{\pi}{2} + n\pi$, $n = 0, 1, 2, \ldots$. The fundamental mode corresponds to $n = 0$. For this mode the wave number and frequency are $k_0 = \pi/(2L)$, $\omega_0 = \pi c/(2L)$. If the nonlinearity given by the right-hand-side of equation (6.57) is taken into account, we have to seek for the solution in the form

$$e = e_0 \sin(\omega t) \cos(k_0 x) \tag{6.59}$$

instead of (6.58). One can see, that the spatial distribution of the mode is the same, but the frequency is different, $\omega \neq \omega_0$. Substituting (6.59) into wave equation (6.57) we derive

$$(-\omega^2 + c^2 k_0^2)e =$$

$$-e^2 \frac{\partial^2}{\partial x^2}[A_1 e_0^2 \sin^2(\omega t)\cos^2(k_0 x) + A_1 A_2 e_0^3 \sin^3(\omega t)\cos^3(k_0 x)] \, (6.60)$$

The quadratic nonlinear term here does not obey the spatial distribution of the $\cos(k_0 x)$ type corresponding to the fundamental mode. Therefore this term cannot influence the solution giving the frequency shift. On the other hand, the cubic nonlinearity contains such a spatial distribution, because

$$\cos^3(k_0 x) = \frac{3}{4}\cos(k_0 x) + \frac{1}{4}\cos(3 k_0 x) \ . \tag{6.61}$$

Keeping the spatial distribution corresponding to the fundamental mode and the temporal vibration with the frequency ω in the cubic nonlinear term, we get the following equation:

$$(-\omega^2 + c^2 k_0^2)e_0 = c^2 k_0^2 [\frac{9}{16} A_1 A_2 e_0^3] \ . \tag{6.62}$$

Consequently, the new nonlinear natural frequency is determined by

$$\omega^2 = \omega_0^2 (1 - \frac{9}{16} A_1 A_2 e_0^2) \ . \tag{6.63}$$

The frequency shift (written for $\frac{9}{16} A_1 A_2 e_0^2$ being small)

$$\frac{f - f_0}{f_0} = -\frac{9}{32} A_1 A_2 e_0^2 \tag{6.64}$$

is negative, i.e. nonlinearity shifts the frequency down. The shift (6.64) is determined by cubic nonlinearity and increases with increase in the amplitude e_0 of vibration. Thus, measuring the amplitude-dependent frequency shift offers the possibility to evaluate the cubic nonlinear module. In turn, the nonlinear elastic modules enable us to evaluate the strength of material.

Problem 6.10

Measurements show that nonlinearity of water containing air bubbles can be very high. Evaluate the maximum possible magnitude of this two-phase medium. Use the equation describing acoustic waves in bubbly water:

$$\frac{\partial^2 p}{\partial x^2} - \frac{1}{c^2}\frac{\partial^2 p}{\partial t^2} = -\rho\nu\frac{\partial^2 w}{\partial t^2} \ , \tag{6.65}$$

and the equation of vibration of a single bubble

$$\frac{\partial^2 w}{\partial t^2} + \omega_0^2 w(1 + \epsilon_a w) = -\omega_0^2\frac{p(t)}{c_a^2\rho_a} \ . \tag{6.66}$$

In the system (6.65) and (6.66) c and ρ are the sound velocity and the density of water, and ϵ_a, c_a and ρ_a are nonlinearity, sound velocity and density of air, $w = V'/V_0$ is the relative disturbance of the equilibrium volume $V_0 = 4\pi R_0^3/3$ of the bubble, and

$$\omega_0^2 = \frac{3c_a^2}{R_0^2}\frac{\rho_a}{\rho} \tag{6.67}$$

is the resonant frequency of bubble. Finally, ν is the volume concentration of air equal to the product of volume of one bubble V_0 times the number of bubbles per unit volume of medium. Take into account that the maximum magnitude of nonlinearity was measured at low acoustic frequencies which are small in comparison with the resonance frequency ω_0 (6.67) of the bubble.

Solution For low frequencies one can neglect the second derivative in the equation (6.66) and reduce it to the algebraic equation

$$w(1 - \epsilon_a w) = -\frac{p(t)}{c_a^2\rho_a} \ , \qquad w \approx -\frac{p}{c_a^2\rho_a} + \epsilon_a\Big(\frac{p}{c_a^2\rho_a}\Big)^2 \ . \tag{6.68}$$

After substitution of the expression (6.68) into equation (6.65) we obtain the nonlinear wave equation

$$\frac{\partial^2 p}{\partial x^2} - \frac{1}{c^2}\frac{\partial^2 p}{\partial t^2} = -\rho\nu\frac{\partial^2}{\partial t^2}\Big[-\frac{p}{c_a^2\rho_a} + \epsilon_a\Big(\frac{p}{c_a^2\rho_a}\Big)^2\Big] \ . \tag{6.69}$$

Using the method of slowly varying profile described in Problem 1.5,

$$p = p(\tau = t - x/c_{\text{eff}}, x_1 = \mu x) , \tag{6.70}$$

we simplify (6.69) and derive the first-order nonlinear equation

$$\frac{\partial p}{\partial x} - \frac{\epsilon_{\text{eff}}}{c_{\text{eff}}^3 \rho_{\text{eff}}} p \frac{\partial p}{\partial \tau} = 0 , \qquad \frac{\epsilon_{\text{eff}}}{c_{\text{eff}}^3 \rho_{\text{eff}}} \equiv \epsilon_a c_{\text{eff}} \frac{\rho \nu}{(c_a^2 \rho_a)^2} . \tag{6.71}$$

Equation (6.71) coincides with the simple (Riemann) wave equation (1.13). At the derivation of (6.71) it is required that terms containing the derivative $\partial^2 p/\partial \tau^2$ must disappear. This is possible if the effective velocity is equal to

$$c_{\text{eff}} = \frac{c}{1 + \nu \beta} , \qquad \beta = \frac{c^2 \rho}{c_a^2 \rho_a} . \tag{6.72}$$

Equation (6.72) can be considered as a definition. It was used in Problem 6.6 and can be derived directly from the system (6.65), (6.66). The effective density of medium $\rho_{\text{eff}} = \rho(1 - \nu)$, naturally, is equal to the density of water subtracted by the fraction of density replaced by air. Using the relations (6.71) and (6.72), we calculate the effective nonlinearity of air-water mixture:

$$\frac{\epsilon_{\text{eff}}}{\epsilon_a} = \frac{\beta^2 \nu(1 - \nu)}{(1 + \beta\nu)^2} . \tag{6.73}$$

One can see, that the maximum magnitude of nonlinearity

$$\epsilon_{\text{eff}} = \epsilon_a \frac{\beta^2}{4(\beta + 1)} \tag{6.74}$$

is reached at the air volume concentration of $\nu = (\beta + 2)^{-1}$. A numerical evaluation shows that the effective nonlinearity exceeds the nonlinearity of diatomic gas (air) $K = \epsilon_{\text{eff}}/\epsilon_a \approx 3900$ times and reaches its maximum $\epsilon_{\text{eff}} \approx 4700$ at the very small air concentration of $\nu = 0.7 \cdot 10^{-4}$.

Chapter 7

FOCUSED NONLINEAR BEAMS AND NONLINEAR GEOMETRICAL ACOUSTICS

Problem 7.1

Derive and analyze the linear solution to the KZ equation (see (4.23)) describing a focused Gaussian initial beam. Instead of using the boundary condition (4.29) which corresponds to a bounded beam with a plane wave initial front, use the boundary condition

$$u(x = 0, r, t) = u_0 \, \exp(-\frac{r^2}{a^2}) \sin[\omega(t + \frac{r^2}{2c_0 R_0})] \ , \qquad (7.1)$$

where R_0 is the initial radius of curvature of the wave front. The phase shift which depends on the transverse coordinate r causes focusing or defocusing. A positive R_0 leads to focusing of the wave.

Solution A calculation identical to that performed at the derivation of solution (4.31), leads for the boundary condition (7.1) to the following result:

$$u(x, r, t) = u_0 \, A(x, r) \sin(\omega t - kx + \Psi(x, r)) \ . \qquad (7.2)$$

The normalized amplitude is given by the formula

$$A(x, r) = \frac{1}{f(x)} \exp[\frac{-r^2}{a^2 f^2(x)}] \ , \qquad f(x) = \sqrt{(1 - \frac{x}{R_0})^2 + \frac{x^2}{x_D^2}} \ . \qquad (7.3)$$

The same definition of characteristic diffraction length $x_D = \omega a^2/(2c_0) = ka^2/2$ is used here as earlier in the Problem 4.12. The phase shift is

$$\Psi(x, r) = \arctan \frac{x/x_D}{1 - x/R_0} - \frac{r^2}{a^2} x_D \frac{d}{dx} \ln f(x) + \pi H(x - R_0) \qquad (7.4)$$

where H is the Heaviside unit step function.

The amplitude on the beam axis (at $r = 0$) is given by (7.3): $A(x) = f^{-1}(x)$. It increases during propagation up to the focal area and reaches its maximum magnitude at

$$\frac{x_{max}}{R_0} = \frac{1}{1 + D^2} \ , \qquad D = \frac{R_0}{x_D} \ . \qquad (7.5)$$

If diffraction is weak and focusing dominates, the inequality $x_D \gg R_0$ is valid and the number D is small. For this important case of sharp focusing, the distance of maximum (7.5) is located near the focal point: $x_{max} \approx R_0$. The maximum amplitude equals to

$$A_{max} = \frac{1}{f(x_{max})} = \frac{\sqrt{1 + D^2}}{D} \ . \qquad (7.6)$$

At weak diffraction the maximum amplitude (7.6) is $1/D$ times higher than the amplitude u_0 of the input wave.

The characteristic width of the beam is $af(x)$ in accordance with formula (7.3). It has a minimum at the same distance (7.5) where the amplitude has the maximum. At weak diffraction the minimum width equals to $aD \ll a$. So, near the focus the beam is very narrow, but its width is finite. Let us be reminded that the geometrical acoustics approximation corresponding to $D \to 0$ predicts an infinite amplitude and zero beam width in the focal point.

The phase shift (7.4) on the axis of the beam $\Psi(R_0, r = 0)$ increases from $\Psi(0,0) = 0$ at the input. It reaches $\Psi(R_0, 0) = \pi/2$ at the focus and increases with further propagation, $\Psi(x \to \infty, 0) \to \pi$. A fast growth of this phase shift takes place near the focus; at $D \to 0$, when the diffraction is negligible, the solution (7.4) transforms to the simple formula $\Psi(x, 0) = \pi H(x - R_0)$. It means, that the phase is zero at $x < R_0$, and equals π at $x > R_0$. In the focal point the phase contains a shock.

Problem 7.2

By using the results of Problem 7.1, demonstrate that the broadband signal (the pulse) changes its shape in the focal region, namely, that the diffraction leads to differentiation of the temporal shape in the point $x = R_0, r = 0$.

Solution First of all, we express the center initial temporal shape of the pulse through Fourier transformation:

$$u(0,0,t) = \int_{-\infty}^{\infty} u_0(\omega) \exp(-i\omega t) \, d\omega \ . \tag{7.7}$$

Now it is convenient to rewrite the solution (7.3), (7.4) given by Problem 7.1 for one frequency component, using other notations:

$$u(x,0,t) = \frac{u_0(\omega)}{\sqrt{(1 - \frac{x}{R_0})^2 + (\frac{2c_0 x}{\omega a^2})^2}} \cdot$$

$$\exp[-i(\omega t - \frac{\omega}{c_0}x + \arctan \frac{\frac{2c_0 x}{\omega a^2}}{1 - \frac{x}{R_0}} + \pi H(x - R_0))] \tag{7.8}$$

At $x = R_0$ this solution takes the form

$$u(R_0,0,t) = \frac{\omega a^2}{2c_0 R_0} u_0(\omega) \exp[-i(\omega t - \frac{\omega}{c_0}R_0 + \pi)] \ . \tag{7.9}$$

Integrating over the full frequency range, we get

$$u(R_0,0,t) = \frac{a^2}{2c_0 R_0} \int_{-\infty}^{\infty} (-i\omega) u_0(\omega) \exp[-i\omega(t - \frac{R_0}{c_0})] \, d\omega \ . \tag{7.10}$$

Formula (7.10) can be rewritten with account for the relation

$$\frac{d}{dt} \exp(-i\omega t) = -i\omega \exp(-i\omega t) \tag{7.11}$$

in the form

$$u(R_0,0,t) = \frac{a^2}{2c_0 R_0} \frac{d}{dt} \int_{-\infty}^{\infty} u_0(\omega) \exp[-i\omega(t - \frac{R_0}{c_0})] \, d\omega \ . \tag{7.12}$$

Comparing formulas (7.12) and (7.7), we conclude, that

$$u(R_0,0,t) = \frac{a^2}{2c_0 R_0} \frac{d}{dt} u(0,0,t - R_0/c_0) \ . \tag{7.13}$$

So, the temporal shape of the pulse measured at the focal point and on the axis of beam is determined by the derivative of the initial form of that pulse. If a single monopolar compressive pulse is radiated by the sound source, it is in the focal area converted into a bipolar pulse containing both compression and rarefaction zones of equal area. Consequently, the steady (zero frequency) component disappears in the focal area, because the lower the frequency, the more rapidly it diffracts.

Problem 7.3

By using the nonlinear KZ equation (see (4.23)), derive simplified equations of nonlinear geometrical acoustics valid at high frequencies [6].

Solution Seeking for the solution to the KZ equation in the form

$$u = u(x, r, T = \tau - \Psi(x, r)/c_0) \ , \tag{7.14}$$

we formally obtain the equivalent equation

$$\frac{\partial}{\partial T}[\frac{\partial u}{\partial x} - \frac{\epsilon}{c_0^2}u\frac{\partial u}{\partial T} - \frac{1}{c_0^2}\frac{\partial \Psi}{\partial x}\frac{\partial u}{\partial T}] = Q \ , \tag{7.15}$$

$$Q \equiv \frac{c_0}{2}\Delta_\perp u - \frac{1}{2}\Delta_\perp\Psi\frac{\partial u}{\partial T} - \frac{\partial \Psi}{\partial r}\frac{\partial^2 u}{\partial r\partial T} + \frac{1}{2c_0}(\frac{\partial \Psi}{\partial r})^2\frac{\partial^2 u}{\partial T^2} \ . \tag{7.16}$$

The notation (7.16) is used for the right-hand-side of equation (7.15). The first term in (7.16) does not contain derivatives of u on T; it is small and can therefore be omitted. All other terms containing the derivatives $\partial u/\partial T \approx \omega u$ and $\partial^2 u/\partial T^2 \approx \omega^2 u$ are much greater and must be kept in the high-frequency approximation. Now the equation (7.15), (7.16) can be integrated over T:

$$\frac{\partial u}{\partial x} - \frac{\epsilon}{c_0^2}u\frac{\partial u}{\partial T} - \frac{1}{c_0^2}\frac{\partial \Psi}{\partial x}\frac{\partial u}{\partial T} = -\frac{u}{2}\Delta_\perp\Psi - \frac{\partial u}{\partial r}\frac{\partial \Psi}{\partial r} + \frac{1}{2c_0}\frac{\partial u}{\partial T}(\frac{\partial \Psi}{\partial r})^2 \ . \tag{7.17}$$

Note that equation (7.17) still contains terms of different order. At $\omega \to \infty$ the two linear u-terms containing the derivative $\partial u/\partial T$ will be maximum in its order. These terms form the so-called "eikonal" equation

$$\frac{\partial \Psi}{\partial x} + \frac{1}{2}\left(\frac{\partial \Psi}{\partial r}\right)^2 = 0 \ . \tag{7.18}$$

The remaining terms in equation (7.17) form the so-called "transport" equation

$$\frac{\partial u}{\partial x} - \frac{\epsilon}{c_0^2} u \frac{\partial u}{\partial T} + \frac{\partial u}{\partial r}\frac{\partial \Psi}{\partial r} + \frac{u}{2}\left(\frac{\partial^2 \Psi}{\partial r^2} + \frac{1}{r}\frac{\partial \Psi}{\partial r}\right) \ . \tag{7.19}$$

Equations (7.18) and (7.19) are known as the system of equations of nonlinear geometrical acoustics. These equations can be solved more easily than the initial KZ equation. However, diffraction phenomena cannot be described in this approximation.

Problem 7.4

Consider a sawtooth-shaped wave, of which one period is given by the formula

$$u = -\frac{\omega T}{\pi}A(x,r) \ , \qquad -\pi < \omega T < \pi \ . \tag{7.20}$$

The straight-line section (7.20) is bounded by two shocks located at $T = \pm\pi$. Derive the equation for the amplitude $A(x,r)$ using the transport equation (7.19) in Problem 7.3.

Solution Substituting (7.20) into the transport equation, we derive

$$\frac{\partial A}{\partial x} - \frac{\epsilon\omega}{\pi c_0^2}A^2 + \frac{\partial A}{\partial r}\frac{\partial \Psi}{\partial r} + \frac{A}{2}\left(\frac{\partial^2 \Psi}{\partial r^2} + \frac{1}{r}\frac{\partial \Psi}{\partial r}\right) = 0 \ . \tag{7.21}$$

As distinct from the initial transport equation, the new equation (7.21) does not contain the temporal variable T. It has another remarkable property, namely, using a substitution $B = 1/A$, we can reduce (7.21) to the linear equation [6]

$$\frac{\partial B}{\partial x} + \frac{\partial B}{\partial r}\frac{\partial \Psi}{\partial r} - \frac{B}{2}\left(\frac{\partial^2 \Psi}{\partial r^2} + \frac{1}{r}\frac{\partial \Psi}{\partial r}\right) = \frac{\epsilon\omega}{\pi c_0^2} \ . \tag{7.22}$$

which can be easily solved for many interesting problems.

Problem 7.5

Derive the solution to equations (7.21) and (7.22) in Problem 7.4, describing the amplitude of a focused sawtooth-shaped wave. The exact solution to the eikonal equation (7.18) in Problem 7.3 for a focusing wave is given by formula

$$\Psi = -\frac{r^2}{2R_0(1 - x/R_0)} \, , \tag{7.23}$$

where R_0 is the radius of curvature of the initial wave front.

Solution With account for (7.23), equation (7.22) in Problem 7.4 takes the form

$$\frac{\partial B}{\partial x} - \frac{r}{R_0 - x}\frac{\partial B}{\partial r} + \frac{B}{R_0 - x} = \frac{\epsilon \omega}{\pi c_0^2} \, . \tag{7.24}$$

The simplest way to solve (7.24) is the following. Through introduction of the new variable $\xi = r^2/(R_0 - x)^2$ one can rewrite (7.24) as an ordinary differential equation

$$\frac{\partial B}{\partial x} + \frac{B}{R_0 - x} = \frac{\epsilon \omega}{\pi c_0^2} \, . \tag{7.25}$$

The solution to (7.25)

$$B = (R_0 - x)[C - \frac{\epsilon \omega}{\pi c_0^2} \ln(R_0 - x)] \tag{7.26}$$

must satisfy the boundary condition

$$B(x = 0, r) = \frac{1}{u_0 \Phi(r^2/a^2)} \tag{7.27}$$

where u_0 is the initial amplitude on the axis of the beam, and Φ is an arbitrary function describing the cross-sectional amplitude distribution of the initial beam. After the constant C in (7.26) is determined from the boundary condition (7.27), we get the following expression [6] describing the amplitude of a strongly distorted nonlinear sawtooth-like wave:

$$A = \frac{u_0}{1 - x/R_0} \Phi[\frac{r^2}{a^2(1 - x/R_0)^2}]$$

$$\{1 - \frac{\epsilon\omega}{\pi c_0^2} u_0 R_0 \ln(1 - \frac{x}{R_0}) \Phi[\frac{r^2}{a^2(1 - x/R_0)^2}]\}^{-1} \ . \quad (7.28)$$

This formula is valid only at distances before the focal point, $x < R_0$. Near the focal point the amplitude tends to infinity, the nonlinear absorption is very strong and the nonlinear geometrical acoustic approximation is incorrect.

It is helpful to compare the nonlinear solution (7.28) with the linear one described in Problem 7.1, which takes into account the diffraction phenomena.

Problem 7.6

Analyze the behaviour of the width of a focused beam which has a Gaussian transverse profile at the input $x = 0$.

Solution Let the function describing the initial transverse distribution be

$$\Phi(x = 0, r) = \exp(-r^2/a^2) \ . \quad (7.29)$$

Let us define the width of the beam as the transverse coordinate $r = r_0(x)$ at which the amplitude is half its axial value:

$$A(x, r_0) = \frac{1}{2} A(x, r = 0) \ . \quad (7.30)$$

Applying the condition (7.30) to the solution (7.28) in Problem 7.5, we obtain the equation describing the dependence of characteristic width of the beam on the distance x passed by the wave:

$$\frac{r_0(x)}{a} = (1 - \frac{x}{R_0}) \ln^{1/2}[2 - \frac{\epsilon\omega}{\pi c_0^2} u_0 R_0 \ln(1 - \frac{x}{R_0})] \ . \quad (7.31)$$

This formula contains two factors. The first one, $(1 - x/R_0)$, describes the decrease in the beam width due to linear focusing. The second one describes the increase in the beam width produced by nonlinear absorption. If the linear focusing is switched off, $R_0 \to \infty$, the nonlinear broadening of the beam is described by the more simple formula:

$$\frac{r_0(x)}{a} = \ln^{1/2}[2 + \frac{\epsilon\omega}{\pi c_0^2}u_0 R_0] \ . \tag{7.32}$$

This phenomenon is known in nonlinear acoustics as nonlinear smoothening of the beam or isotropization of the directivity pattern [4].

Problem 7.7

Derive equations of nonlinear geometrical acoustics like (7.18) and (7.19) in Problem 7.3 but for beams which are not round in their cross-section.

Answer Using the KZ equation written in a Cartesian coordinate system, one can derive the following eikonal and transport equations:

$$\frac{\partial\Psi}{\partial x} + \frac{1}{2}(\frac{\partial\Psi}{\partial y})^2 + \frac{1}{2}(\frac{\partial\Psi}{\partial z})^2 = 0 \ . \tag{7.33}$$

$$\frac{\partial u}{\partial x} - \frac{\epsilon}{c_0^2}u\frac{\partial u}{\partial T} + \frac{\partial u}{\partial y}\frac{\partial\Psi}{\partial y} + \frac{\partial u}{\partial z}\frac{\partial\Psi}{\partial z} + \frac{u}{2}(\frac{\partial^2\Psi}{\partial y^2} + \frac{\partial^2\Psi}{\partial z^2}) = 0 \ . \tag{7.34}$$

We remind that the x-axis coincides with the direction of wave propagation, and the two other coordinates y and z are introduced in the cross-section of the beam.

Problem 7.8

Rewrite the system of eikonal and transport equations using new variables $\alpha = \partial\Psi/\partial y$, $\beta = \partial\Psi/\partial z$. Because $\Psi = constant$ determines the wave front (the surface of equal phase) and $\vec{n} = \nabla\Psi$ is the vector normal to this surface (rays are parallel to $\vec{n}$), the variables α and β have a simple geometrical meaning - they are cosines of the ray inclination to the y and x axes correspondingly. Solve the eikonal equation rewritten through α and β.

Solution Equation (7.33) in Problem 7.7, can be reduced to the system

$$\frac{\partial\alpha}{\partial x} + \alpha\frac{\partial\alpha}{\partial y} + \beta\frac{\partial\alpha}{\partial z} = 0 \ , \qquad \frac{\partial\beta}{\partial x} + \alpha\frac{\partial\beta}{\partial y} + \beta\frac{\partial\beta}{\partial z} = 0 \qquad (7.35)$$

by successive differentiation of the eikonal equation on variables y and z. The transport equation takes form

$$\frac{\partial u}{\partial x} - \frac{\epsilon}{c_0^2}u\frac{\partial u}{\partial T} + \alpha\frac{\partial u}{\partial y} + \beta\frac{\partial u}{\partial z} + (\frac{\partial\alpha}{\partial y} + \frac{\partial\beta}{\partial z})\frac{u}{2} = 0 \ . \qquad (7.36)$$

The general solution to the system (7.35) is given by the two-dimensional implicit functions .

$$\alpha = A(\xi = y - \alpha x, \eta = z - \beta x) \ , \qquad \beta = B(\xi = y - \alpha x, \eta = z - \beta x) \ . \qquad (7.37)$$

Here the arbitrary functions A and B are connected to each other by the differential relation $\partial A/\partial z = \partial B/\partial z$ following from the definition of ray inclinations as derivatives of the eikonal.

Problem 7.9[5]

Transform equations (7.35) and (7.36) in Problem 7.8 to the Riemann wave equation using transition from y and z to the new Lagrangian variables $\xi = y - \alpha x$, $\eta = z - \beta x$.

Solution Change of variables lead to the following transformations of derivatives:

$$\frac{\partial\alpha}{\partial y} = \frac{1}{S}[\frac{\partial A}{\partial\xi} + x(\frac{\partial A}{\partial\xi}\frac{\partial B}{\partial\eta} - \frac{\partial A}{\partial\eta}\frac{\partial B}{\partial\xi})]$$

$$\frac{\partial\beta}{\partial z} = \frac{1}{S}[\frac{\partial B}{\partial\eta} + x(\frac{\partial A}{\partial\xi}\frac{\partial B}{\partial\eta} - \frac{\partial A}{\partial\eta}\frac{\partial B}{\partial\xi})]$$

---

[5]This problem follows the paper: V.A.Gusev, O.V.Rudenko. Acoustical Physics **2**, 2006

$$\frac{\partial \alpha}{\partial z} = \frac{\partial \beta}{\partial y} = \frac{1}{S}\frac{\partial A}{\partial \eta} = \frac{1}{S}\frac{\partial B}{\partial \xi}$$

$$S = 1 + x\left(\frac{\partial A}{\partial \xi} + \frac{\partial B}{\partial \eta}\right) + x^2\left(\frac{\partial A}{\partial \xi}\frac{\partial B}{\partial \eta} - \frac{\partial A}{\partial \eta}\frac{\partial B}{\partial \xi}\right) \ . \tag{7.38}$$

The function S is the Jacobian of transformation from old variables to new ones. This function has a transparent geometrical sense; it equals the elementary cross-section area of a ray tube formed by neighbouring rays:

$$dy\,dz = \frac{D(y, z)}{D(\xi, \eta)}d\xi\,d\eta \equiv S \cdot d\xi\,d\eta \ . \tag{7.39}$$

Now it is necessary to replace the derivatives in the equation (7.36) in Problem 7.8 by their new expressions (7.38). So, now the acoustic particle velocity is a function of new variables: $u = u(x, \xi, \eta, T)$. It follows from formulas (7.38), that the last term in the transport equation (7.36) in Problem 7.8, now equals

$$\left(\frac{\partial \alpha}{\partial y} + \frac{\partial \beta}{\partial z}\right)\frac{u}{2} \quad \Rightarrow \quad \frac{u}{2}\frac{\partial}{\partial x}\ln S(x, \xi, \eta) \ , \tag{7.40}$$

Using formulas for other derivatives

$$\frac{\partial \xi}{\partial x} = -\frac{1}{S}\left[A + x\left(A\frac{\partial B}{\partial \eta} - B\frac{\partial A}{\partial \eta}\right)\right] \ ,$$

$$\frac{\partial \eta}{\partial x} = -\frac{1}{S}\left[B + x\left(B\frac{\partial A}{\partial \xi} - A\frac{\partial B}{\partial \xi}\right)\right] \ ,$$

$$\frac{\partial \xi}{\partial y} = \frac{1}{S}\left(1 + x\frac{\partial B}{\partial \eta}\right) \ , \qquad \frac{\partial \xi}{\partial z} = -\frac{x}{S}\frac{\partial A}{\partial \eta} \ ,$$

$$\frac{\partial \eta}{\partial y} = -\frac{x}{S}\frac{\partial B}{\partial \xi} \ , \qquad \frac{\partial \eta}{\partial z} = \frac{1}{S}\left(1 + x\frac{\partial A}{\partial \xi}\right) \ , \tag{7.41}$$

one can transform the remaining linear terms in the transport equation to the simple form

$$\frac{\partial u}{\partial x} + \alpha\frac{\partial u}{\partial y} + \beta\frac{\partial u}{\partial z} \quad \Rightarrow \quad \frac{\partial}{\partial x}u(x, \xi, \eta, T) \ . \tag{7.42}$$

As a result of these transformations, the transport equation takes the form of a Riemann-type wave equation

$$\frac{\partial u}{\partial x} + \frac{u}{2}\frac{d}{dx}\ln S - \frac{\epsilon}{c_0^2}u\frac{\partial u}{\partial T} = 0 \ . \tag{7.43}$$

The simplicity of the equation (7.43) which is similar to the ordinary Riemann wave equation (see (1.13), Problem 1.5) is in its one-dimensional character. Really, the derivatives in (7.43) contain only two variables x and T. Two other Lagrangian variables ξ and η play the role of parameters; they are the same for the given ray for any distance x passed by the wave through the nonlinear medium.

Problem 7.10

Show that Webster's equation containing an additional nonlinear term

$$\frac{\partial^2 u}{\partial x^2} + \frac{\partial u}{\partial x}\frac{d}{dx}\ln S(x) - \frac{1}{c_0^2}\frac{\partial^2 u}{\partial t^2} = -\frac{\epsilon}{c_0^3}\frac{\partial^2 u^2}{\partial t^2} \tag{7.44}$$

can be reduced to equation (7.43) in Problem 7.9. Describe the physical situation when such simplification is possible. It is known, that the linear Webster equation is used to describe waves in tubes with a cross-section area depending on the coordinate along the axis of the tube, in particular in horns (increasing $S(x)$ increases) and concentrators (decreasing $S(x)$).

Solution Through following the general scheme of the slowly varying profile method, which is described in Problem 1.5, we seek the solution to (7.44) in the form

$$u = u(\tau = t - x/c_0, x_1 = \mu x) \ . \tag{7.45}$$

Evidently formula (7.45) corresponds to a wave traveling along the x-axis. A wave propagating in the opposite direction, as well as a standing wave cannot be described by this approach. Moreover, the function must be slowly varying on the scale of the wave length. Under such conditions the following simplified expressions for derivatives in (7.44) are valid:

$$\frac{\partial^2 u}{\partial x^2} \approx \frac{1}{c_0^2}\frac{\partial^2 u}{\partial \tau^2} - \frac{2}{c_0}\mu\frac{\partial^2 u}{\partial \tau \partial x_1} , \qquad \frac{\partial u}{\partial x}\frac{d}{dx}\ln S(x) \approx -\frac{1}{c_0}\frac{\partial u}{\partial \tau}\mu\frac{d}{dx_1}\ln S(x) ,$$

$$\frac{1}{c_0^2}\frac{\partial^2 u}{\partial t^2} = \frac{1}{c_0^2}\frac{\partial^2 u}{\partial \tau^2} , \qquad -\frac{\epsilon}{c_0^3}\frac{\partial^2 u^2}{\partial t^2} = -\frac{\epsilon}{c_0^3}\frac{\partial^2 u^2}{\partial \tau^2} . \tag{7.46}$$

Substituting the new terms (7.46) into (7.44), we have derived equation (7.43) in Problem 7.9.

Problem 7.11

Simplify equation (7.43) in Problem 7.9, using the new variables

$$V = \frac{u}{u_0}\sqrt{\frac{S(x)}{S(0)}} , \qquad \theta = \omega\tau , \qquad z = \frac{\epsilon\omega}{c_0^2}u_0\int_0^x \sqrt{\frac{S(0)}{S(x')}}\,dx' . \tag{7.47}$$

Answer In the new variables equation (7.43), Problem 7.9, takes the simplest form

$$\frac{\partial V}{\partial z} - V\frac{\partial V}{\partial \theta} = 0 . \tag{7.48}$$

Its solution corresponding to the wave at $z = 0$ being $V(z = 0, \theta) = \Phi(\theta)$, is

$$V = \Phi(\theta + zV) , \qquad u = u_0\sqrt{\frac{S(0)}{S(x)}}(\omega\tau + \frac{\epsilon\omega}{c_0^2}u\int_0^x \sqrt{\frac{S(0)}{S(x')}}\,dx') . \tag{7.49}$$

Problem 7.12

Add a dissipative term to the equation (7.43) in Problem 7.9. By using the variables (7.47) from Problem 7.11, reduce the modified equation to the canonical form of Generalized Burgers equation (GBE)[6].

---

[6]See details in the paper: B.O.Enflo, O.V.Rudenko. Acta Acustica **88**, 155-162, 2002

Answer With account for dissipation, equation (7.43) takes the form

$$\frac{\partial u}{\partial x} + \frac{u}{2}\frac{d}{dx}\ln S - \frac{\epsilon}{c_0^3}u\frac{\partial u}{\partial T} = \frac{b}{2c_0^3\rho_0}\frac{\partial^2 u}{\partial T} \ . \tag{7.50}$$

It reduces to the Generalized Burgers equation:

$$\frac{\partial V}{\partial z} - V\frac{\partial V}{\partial \theta} = \Gamma(z)\frac{\partial^2 V}{\partial \theta^2} \ , \qquad \Gamma(z) = \Big[\frac{b\omega}{2\epsilon c_0\rho_0 u_0}\sqrt{\frac{S(x)}{S(0)}}\Big]_{x=x(z)} \ . \tag{7.51}$$

Here $\Gamma(z)$ is the coordinate-dependent inverse acoustical Reynolds number. For a plane wave it does not depend on the path length traversed by the wave (see (3.11) in Problem 3.2).

Problem 7.13

Show that for focused spherical and cylindrical waves the General Burgers equation (GBE) can be reduced to the equation (4.7) in Problem 4.3 (spherically converging wave), and to the equation (4.10) in Problem 4.5 (cylindrically converging wave).

Solution For spherical focusing the cross-section area of a ray tube decreases according to the law

$$S(x) = (1 - \frac{x}{R_0})^2 \ , \tag{7.52}$$

where R_0 is the distance from the source to the focal point. The corresponding dimensionless coordinate is

$$z = \frac{\epsilon}{c_0^2}\omega u_0 \int_0^x \frac{dx'}{1 - x'/R_0}\, dx' = -z_0 \ln(1 - \frac{x}{R_0}) \ , \qquad z_0 = \frac{\epsilon}{c_0^2}\omega u_0 R_0 \ . \tag{7.53}$$

Using (7.47) in Problem 7.11, we calculate

$$\Gamma(z) = \Big[\frac{b\omega}{2\epsilon c_0\rho_0 u_0}(1 - \frac{x}{R_0})\Big]_{x=x(z)} = \Gamma\exp(-\frac{z}{z_0}) \ , \qquad \Gamma \equiv \frac{b\omega}{2\epsilon c_0\rho_0 u_0} \ . \tag{7.54}$$

Its corresponding GBE is

$$\frac{\partial V}{\partial z} - V \frac{\partial V}{\partial \theta} = \Gamma \exp(-\frac{z}{z_0}) \frac{\partial^2 V}{\partial \theta^2} \ . \tag{7.55}$$

It coincides with equation (4.7), with a difference only in notations. By analogy, using the formula for the cross-section area of a focused cylindrical wave

$$S(x) = (1 - \frac{x}{R_0}) \ , \tag{7.56}$$

we calculate ,

$$z = -2z_0[\sqrt{1 - x/R_0} - 1] \ , \quad \sqrt{S[(x(z)]} = 1 - z/(2z_0) \ , \tag{7.57}$$

and

$$\frac{\partial V}{\partial z} - V \frac{\partial V}{\partial \theta} = \Gamma(1 - \frac{z}{2z_0}) \frac{\partial^2 V}{\partial \theta^2} \ . \tag{7.58}$$

This equation has the same form as equation (4.10).

Problem 7.14

Derive the GBE for an exponential concentrator where the cross-section area decreases according to the law

$$S(x) = S(0) \exp(-x/R_0) \ . \tag{7.59}$$

Obtain its exact solution using separation of variables.

Solution Following the approach used at the derivation of GBE for spherical and cylindrical waves in Problem 7.13, we derive

$$\frac{\partial V}{\partial z} - V \frac{\partial V}{\partial \theta} = \Gamma(1 + \frac{z}{z_0})^{-1} \frac{\partial^2 V}{\partial \theta^2} \ . \tag{7.60}$$

This equation has a solution with separated variables:

$$V(z, \theta) = f(\theta)(1 + z/z_0)^{-1} \ , \tag{7.61}$$

where the function $f(\theta)$ satisfies the ordinary differential equation

$$\Gamma f'' + f f' + z_0^{-1} f = 0 \ . \tag{7.62}$$

Equation (7.62) has a periodic solution in form of a sawtooth-like wave. At weak dissipation (small Γ-numbers) one period of this wave is given by the asymptotic formula of the same type as the well-known Khokhlov solution (3.45) in Problem 3.14:

$$f(\theta) = \frac{1}{z_0}[-\theta + \pi \tanh(\frac{\pi\theta}{2\Gamma z_0})] \ . \tag{7.63}$$

So, the solution (7.61) has a stable temporal profile (7.63), but its amplitude decreases according to the law $(1 + z/z_0)^{-1}$ because of nonlinear absorption. The steepness of the shock front does not depend on the distance because of competition between two processes: the nonlinear absorption and the nonlinear steepening. These processes are in a state of dynamic equilibrium when the wave converges inside the exponential concentrator.

Problem 7.15

An acoustical beam heats the medium in which it propagates because of loss of part of its energy. The temperature increases inside the beam and the medium gets lens-like properties. This medium is now inhomogeneous in the transverse direction and can focus or defocus the beam. The heating is stronger near the axis, as is the deviation in sound velocity. The eikonal equation (7.18) in Problem 7.3

$$\frac{\partial\Psi}{\partial z} + \frac{1}{2}(\frac{\partial\Psi}{\partial r})^2 = -\delta T \tag{7.64}$$

will now contain a righthand-side where T is the increase in temperature caused by the acoustic beam, and δ is the temperature coefficient of the sound velocity. Taking the temperature distribution near the axis in parabolic form $T(r) = T_0(1 - r^2/a^2)$, calculate the form of rays for positive and negative values of the coefficient δ.

Solution The solution to the eikonal equation at parabolic temperature distribution is sought in the following form:

$$\Psi(r, x) = \Psi_0(x) + (r^2/2)f(x) .\tag{7.65}$$

Substituting (7.65) into (7.64), we get a pair of equations for the unknown functions

$$\frac{d\Psi_0}{dx} = \delta T_0 , \quad \frac{df}{dx} + f^2 = \frac{2\delta T_0}{a^2} \equiv \pm E^2 .\tag{7.66}$$

The '+'-sign corresponds here to a positive δ, and the '−'-sign corresponds to a negative coefficient δ. Solving (7.66) with the boundary condition $\Psi(x = 0, r) = 0$, we calculate the eikonal function for these two cases. For positive δ

$$\Psi(r, x) = \delta T_0 x + (r^2/2)|E|\tanh x ,\tag{7.67}$$

and for negative δ

$$\Psi(r, x) = \delta T_0 x - (r^2/2)|E|\tan x .\tag{7.68}$$

The cosine of the ray inclination angle to the r-axis is $\partial\Psi/\partial r$. For small angles the ray is almost parallel to x-axis, and the corresponding cosine is equal to unity. The equation describing acoustic rays inclined at small angles to the axis of the beam, is

$$\frac{dx}{1} = \frac{dr}{\partial\Psi/\partial r} .\tag{7.69}$$

Integrating this equation, for positive δ (7.67) we get the following description of rays

$$r = r_0 \cosh\left(x\sqrt{\frac{2\delta T_0}{a^2}} \right) .\tag{7.70}$$

The form of rays (7.70) is shown in Figure (7.1a) for two values of temperatures T_0. This case corresponds to a defocusing medium, the most typical example of this is water at room temperature. For negative δ (7.68) the form of rays

$$r = r_0 \cos\left(x\sqrt{\frac{2\delta T_0}{a^2}} \right) ,\tag{7.71}$$

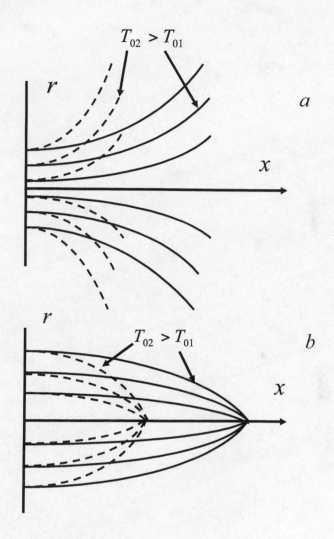

Figure 7.1: *Rays in focusing (a) and defocusing (b) media.*

is shown in Figure (7.1b).

This case corresponds to focusing media (examples are acetone or alcohol). In both formulas the temperature T_0 depends on the sound intensity. These phenomena are known as self-focusing and self-defocusing. A detailed description of these processes is given in the review by O.V.Rudenko, O.A.Sapozhnikov, Physics-Uspekhi (Advances in Physical Sciences) 47(9), 2004.

Chapter 8

VARIOUS TYPES OF NONLINEAR PROBLEMS

Problem 8.1[7]

Let a plane piston vibrate with very high amplitude along the normal to its surface which coincides with the x-axis. The displacement of the piston from its equilibrium position at $x = 0$ is described by the known function $x = X(t)$. Determine the temporal shape and spectrum of acoustic wave excited by this piston in the medium. The wave propagates in positive direction of the x-axis, $x > 0$, and the particle velocity varies as $u = u_0 \Phi(t - x/c)$ where Φ is an unknown function.

Solution Common opinion has it that the form of the particle velocity wave is the same as the form of the velocity of piston, because

$$u(x = 0, t) = u_0 \Phi(t) = dX/dt \ . \tag{8.1}$$

However, the conclusion (8.1) is incorrect. The coordinate of the piston is not zero, it is $x = X(t)$ and the assumption that the boundary condition is satisfied at $x = 0$ at any moment of time is approximately valid only at small amplitudes of vibration. The correct boundary condition must be satisfied on the surface of piston, and instead of (8.1) it has the form

$$u(x = X(t), t) = u_0 \Phi(t - X(t)/c) = dX/dt \ . \tag{8.2}$$

This is a functional equation for determination of the unknown function Φ. Using the approach described in Problems 1.12 and 1.19 at the

---

[7]Problems 8.1 and 8.2 are based on results of paper: O.V.Rudenko, Acoustical Physics **44**(6), 717, 1998

calculation of spectra of an implicit function (of Bessel-Fubini type), we can write the solution to (8.2) in integral form:

$$\Phi(t) = \frac{1}{2\pi i} \int_{-\infty}^{\infty} \exp(i\omega t) \frac{d\omega}{\omega} \int_{-\infty}^{\infty} \frac{d^2 X}{d\xi^2} \exp[i\frac{\omega}{c} X(\xi) - i\omega\xi] \, d\xi \ . \quad (8.3)$$

Formula (8.3) demonstrates the nonlinear dependence of the wave profile on the known function $X(t)$ describing the piston vibration.

Problem 8.2

Using the general formula (8.3), derived in Problem 8.1, calculate the profile and spectrum of the wave radiated by a harmonically vibrating piston by use of the formula $X(t) = -X_0 \cos(\omega t)$.

Solution The functional equation for a harmonic vibration has the form

$$\frac{u}{u_0} = \Phi(\omega t + \frac{u_0}{c} \cos(\omega t)) = \sin(\omega t) \ , \qquad u_0 = \omega X_0 \ . \quad (8.4)$$

Substituting the given function $X(\xi)$ into the integral (8.3) we derive a series expansion in which the nth term corresponds to the nth harmonic:

$$\frac{u}{u_0} = \Phi(\omega t) = \sum_{n=0}^{\infty} \frac{i^{-n}}{n} J'(nM) \cdot \exp(in\omega t) =$$

$$\sum_{n=0}^{\infty} (A_n \cos(n\omega t) + B_n \sin(n\omega t)) \ . \quad (8.5)$$

In particular, the following relations for the steady-state component and the two first harmonics are valid:

$$A_0 = -\frac{M}{2} \ , \qquad B_1 = J_0(M) - J_2(M) \ , \qquad A_2 = -\frac{1}{2}[J_1(2M) - J_3(2M)] \ . \quad (8.6)$$

It is helpful to compare the result (8.5) with the Bessel-Fubini formula (1.40) in Problem 1.13.

The series (8.5) contains both a steady-state component and higher harmonics, despite the fact that $X(t)$ is a harmonic function. The difference between $X'(t)$ and $\Phi(t)$ increases with increase in the Mach number $M = u_0/c$. The nonlinear distortion caused by the movable boundary nonlinearity is well pronounced at velocities of piston vibration approaching the sound velocity. Such nonlinear phenomena can be observed in air-saturated water where sound velocity is very small (it falls down to 24 m/s — much lower than the velocities in both liquid and gas). Another system is a high-Q resonator with a vibrating wall where nonlinearity can accumulate considerably with time. The form of one wave period (8.3) is shown in Figure (8.1). At small Mach numbers the profile is close to harmonic, but at finite Mach numbers it is non-symmetric: the section of positive velocity is shorter than the negative half-period, and the positive peak is higher than the absolute value of the negative peak. As distinct from the volume nonlinearity

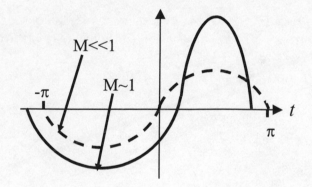

Figure 8.1: *Profile of one period of a traveling wave excited by harmonic vibration of the piston. Distortion appears because of the boundary nonlinearity.*

considered in other problems in this book, the nonlinearity analyzed in Problems 8.1, 8.2 can be referred to as boundary geometrical nonlinearity, because it depends neither on the physical properties of the medium, nor on the distance passed by the wave. Its manifestation becomes stronger with increase in Mach number.

Problem 8.3

One-dimensional vibrations inside a resonator can be described by the nonlinear wave equation written for the particle velocity:

$$\frac{\partial^2 u}{\partial x^2} - \frac{1}{c_0^2}\frac{\partial^2 u}{\partial t^2} = -\frac{\epsilon}{c^3}\frac{\partial^2 u^2}{\partial t^2} \ . \tag{8.7}$$

The left boundary vibrates harmonically, whereas the right-hand side is immovable:

$$u(x = 0, t) = A\sin(\omega t) \ , \qquad u(x = L, t) = 0 \ . \tag{8.8}$$

Consider first linear vibrations using equation (8.7) with $\epsilon = 0$ and boundary conditions (8.8). Analyze both non-resonant and resonant excitation.

Solution It is known that the general solution of the linear wave equation consists of two counter-propagating waves of arbitrary shape:

$$u(x, t) = F_+(t - \frac{x - L}{c}) + F_-(t + \frac{x - L}{c}) \ . \tag{8.9}$$

Using the second boundary condition (8.8) at $x = L$, we determine that the forms of the two waves are the same but they have opposite signs: $F_+(t) = -F_-(t) \equiv F(t)$. Satisfying the boundary condition at the vibrating wall, we derive a functional equation:

$$F(\omega t + kL) - F(\omega t - kL) = A\sin(\omega t) \ . \tag{8.10}$$

As distinct from differential equations, functional equations are usually more complicated and rarely have analytical solutions. However, the equation (8.10) has a general solution which is a sum of the partial solution of the inhomogeneous equation and the general solution of the homogeneous equation:

$$F = \frac{A\cos(\omega t)}{2\sin(kL)} + \sum_{n=0}^{\infty}[A_n\cos(n\omega_0 t) + B_n\sin(n\omega_0 t)] \ . \tag{8.11}$$

The natural frequency of the fundamental mode is determined by

$$\omega_0 = \pi c/L \ , \qquad L = \lambda_0/2 \ . \tag{8.12}$$

One can easily check by direct substitution that expression (8.11) satisfies the functional equation (8.10) and both boundary conditions (8.8).

Let us now consider resonant vibrations. We assume that the frequency of vibration of the wall approaches one of the natural frequencies, say, the frequency of the fundamental mode, $\omega \to \omega_0$. In this case, an unsteady-state resonant growth of amplitude takes place. To demonstrate this phenomenon, choose the coefficients in the general solution (8.11) to be

$$A_1 = \frac{A}{2\sin(kL)} \ , \qquad B_n = 0 \ , \qquad A_n = 0 \ (n \neq 1) \ . \tag{8.13}$$

The solution (8.11) with account for (8.13) contains a singularity of the "0/0" type. Resolving this singularity, we get the resonant solution:

$$F = \frac{A}{2} \lim_{\omega \to \omega_0} \frac{\cos(\omega_0 t) - \cos(\omega t)}{\sin(\omega L/c)} = \frac{A}{2\pi}(\omega_0 t)\sin(\omega_0 t) \ . \tag{8.14}$$

It describes a harmonic vibration where the amplitude grows unrestrictedly with time. This growth is usually limited by absorption or by nonlinear phenomena.

Problem 8.4

In Problem 8.3 the linear standing wave is composed of two plane waves propagating in opposite directions. This representation can be generalized for nonlinear standing waves, where the vibration is described as a sum of two Riemann or Burgers traveling waves. These waves can be distorted significantly by nonlinear self-action, resulting in the formation of a sawtooth-like profile from the initial harmonic, with no contribution from the cross-interaction. In other words, each wave is distorted by itself during the propagation — there is no significant energy exchange between them. Show that nonlinear standing waves can be represented as a linear sum of two counter-propagating nonlinear waves. Use the successive approximation method to derive a weakly nonlinear solution to equation (8.7) in Problem 8.3.

Solution Equation (8.7) govern Riemann waves propagating in opposite directions, as well as the interaction between them. The solution is sought by successive approximation (see Problem 1.3):

$$u = u^{(1)} + u^{(2)} + \dots \ . \tag{8.15}$$

Let the linear (first) approximation solution be a sum of two counter-propagating harmonic waves :

$$u^{(1)} = B_1 \cos(\omega_1 t - k_1 x) + B_2 \cos(\omega_2 t + k_2 x) \ , \quad k_{1,2} = \omega_{1,2}/c \ . \tag{8.16}$$

The second approximation is derived from the inhomogeneous linear equation

$$\frac{\partial^2 u^{(2)}}{\partial x^2} - \frac{1}{c_0^2}\frac{\partial^2 u^{(2)}}{\partial t^2} = F(2\omega_1) + F(2\omega_2) + F(\omega_1 + \omega_2) + F(\omega_1 - \omega_2) \ , \tag{8.17}$$

whose right-hand side is calculated on the basis of the linear approximation solution (8.16):

$$F(2\omega_{1,2}) = \frac{2\epsilon}{c^3}\omega_{1,2}^2 \cdot B_{1,2}^2 \cos[2(\omega_{1,2}t \mp k_{1,2}x)]$$

$$F(\omega_1 \pm \omega_2) = \frac{\epsilon}{c^3}(\omega_1 \pm \omega_2)^2 \cdot B_1 B_2 \cos[(\omega_1 \pm \omega_2)t - (k_1 \mp k_2)x)] \tag{8.18}$$

In the context of the approximate method (8.15) the four terms in the right-hand side of equation (8.17) can be considered as external forces exciting the second approximation forced waves at frequencies of second harmonics, $2\omega_1$ and $2\omega_2$, as well as at sum $(\omega_1 + \omega_2)$ and difference $(\omega_1 - \omega_2)$ frequencies.

It is important that the excitation of secondary waves can have either a resonant character or a non-resonant one. The first two forces $F(2\omega_1)$ and $F(2\omega_2)$ lead to resonant excitation. The corresponding forced waves

$$u_{1,2}^{(2)} = -\frac{\epsilon}{2c}(\omega_{1,2}t)B_{1,2}^2 \sin[2(\omega_{1,2}t \mp k_{1,2}x)] \tag{8.19}$$

increase with time. The amplitudes grow linearly like the amplitude of forced vibrations at coincidence of the natural and driving frequencies (see (8.14) in Problem 8.3).

The partial solution to equation (8.17) corresponding to the other two forced waves excited by the $F(\omega_1 \pm \omega_2)$ forces in (8.18) is

$$u_{3,4}^{(2)} = \frac{\epsilon}{c^3} \frac{(\omega_1 \pm \omega_2)^2}{4k_1 k_2} B_1 B_2 \cos[(\omega_1 \pm \omega_2)t - (k_1 \mp k_2)x)] \ . \qquad (8.20)$$

Its amplitudes are independent of t.

The comparison between the resonant (8.19) and non-resonant (8.20) solutions shows that after several periods of vibration, $(\omega_{1,2}t) \gg 1$, the waves (8.20) become much weaker than the resonant ones (8.19), and cannot participate significantly in the nonlinear energy exchange. Consequently, each of two waves propagating in opposite directions generates its higher harmonics (8.19), but the cross-interaction processes (8.20) can be neglected if the waves are periodic in time. This conclusion is easily seen to be valid for periodic waves intersecting at any sufficiently large angles depending on the acoustic Mach number [4].

The resulting linear superposition of two nonlinear Riemann waves which generalizes the functional equation (8.18) is

$$F[\omega t + kL - \frac{\epsilon}{c}kLF] - F[\omega t - kL + \frac{\epsilon}{c}kLF] = A\sin(\omega t) \ . \qquad (8.21)$$

Unfortunately, the nonlinear functional equations of implicit form like (8.21) are complicated and cannot be solved analytically. It is necessary to use approximative methods to analyze the behaviour of nonlinear standing waves.

Problem 8.5

The nonlinear functional equation (8.21) in Problem 8.4 can be reduced to a simplified evolution equation, if the length of resonator is small in comparison with the nonlinear length, and if the frequency of vibration of the left boundary differs only slightly from the resonant frequency:

$$L \ll \frac{c^2}{\epsilon \omega F_{max}} , \quad kL = \pi + \Delta , \quad \Delta = \pi \frac{\omega - \omega_0}{\omega_0} \ll 1 . \quad (8.22)$$

Here F_{max} is the maximum value of the function F, and Δ is a frequency discrepancy. The second condition in (8.22) corresponds to the fundamental resonance ($n = 1$).

Derive the simplified nonlinear differential equation.

Solution With the restrictions in (8.22), the left-hand side of equation (8.21) in Problem 8.4 can be expanded in a series:

$$F[\omega t + \pi + \Delta - \pi \frac{\epsilon}{c} F] - F[\omega t - \pi - \Delta + \pi \frac{\epsilon}{c} F] \quad (8.23)$$

$$\approx [F(\omega t + \pi) - F(\omega t - \pi)] + (\Delta - \pi \frac{\epsilon}{c} F)[F'(\omega t + \pi) + F'(\omega t - \pi)] .$$

It is evident that F is a quasi-periodic function, whose parameters are slowly varying in time. Therefore,

$$F(\omega t + \pi) - F(\omega t - \pi) \approx 2\pi\mu \frac{\partial F}{\partial(\mu\omega t)} , \quad (8.24)$$

where $\mu \ll 1$ is a small parameter whose physical meaning will be clear later. Equation (8.24) will have the form

$$\mu \frac{\partial F}{\partial(\frac{\mu}{\pi}\omega t)}) + (\Delta - \pi \frac{\epsilon}{c} F) \frac{\partial F(\omega t + \pi)}{\partial(\omega t)} = \frac{A}{2} \sin(\omega t) . \quad (8.25)$$

Introducing new dimensionless variables and constants

$$\xi = \omega t + \pi , \quad U = \frac{F}{c} , \quad M = \frac{A}{c} , \quad T = \frac{\omega t}{\pi} , \quad (8.26)$$

one can rewrite equation (8.25) as

$$\frac{\partial U}{\partial T} + \Delta \frac{\partial U}{\partial \xi} - \pi\epsilon U \frac{\partial U}{\partial \xi} = \frac{M}{2} \sin \xi . \quad (8.27)$$

Two temporal variables are introduced in equation (8.27); ξ is the "fast" time responsible for oscillation, and T is the "slow" time describing the evolution of the wave. It is now evident that the small parameter μ can play the role of any of the small numbers: Δ, M, or $U \sim M$. Equation (8.27) is known as the "Inhomogeneous Riemann wave equation with the discrepancy" [6].

Problem 8.6

Derive a solution to the Inhomogeneous Riemann wave Equation (IRE) (8.27) in Problem 8.5, putting nonlinearity to zero.

Solution The equation for the characteristics of the partial differential equation IRE (for $\epsilon = 0$) is

$$\frac{dT}{1} = \frac{d\xi}{\Delta} = \frac{dU}{(M/2)\sin\xi} . \tag{8.28}$$

The solution to this system is

$$\xi - \Delta T = C_1 , \qquad U + \frac{M}{2\Delta}\cos(C_1 + \Delta T) = C_2 . \tag{8.29}$$

Consider the following initial condition: there is no vibration at $T = 0$, i.e. $U(T = 0) = 0$. The partial solution, corresponding to this initial condition, has the form

$$C_2 = \frac{M}{2\Delta}\cos C_1$$

$$\frac{U}{M} = \frac{\cos(\xi - T\Delta) - \cos\xi}{2\Delta} = \frac{1}{\Delta}\sin(\frac{\Delta}{2}T)\sin(\xi - \frac{\Delta}{2}T) . \tag{8.30}$$

One can see that the wave amplitude oscillates as

$$|U| = \frac{M}{\Delta}\sin(\frac{\Delta}{2}T) . \tag{8.31}$$

With increase in the discrepancy $|\Delta|$, both the period of oscillations and the maximum amplitude decrease. If the discrepancy tends to zero, we have a linear resonant growth proportional to time:

$$|U| = \frac{M}{2}T , \qquad U = \frac{M}{2}T\sin\xi . \tag{8.32}$$

Problem 8.7

Derive a steady-state solution to the Inhomogeneous Riemann wave Equation (IRE) (8.27) in Problem 8.5, obtained for $T \to \infty$.

Solution The steady-state equation does not depend on the slow time

$$\Delta \frac{dU}{d\xi} - \pi\epsilon U \frac{dU}{d\xi} = \frac{M}{2}\sin\xi \ . \tag{8.33}$$

After integration over ξ we get

$$-2U\Delta + \pi\epsilon U^2 = M(\cos\xi + C) \ . \tag{8.34}$$

The algebraic equation (8.34) must satisfy an additional condition; the mean value of the acoustic velocity averaged over one period must be equal to zero (no streaming):

$$\overline{U} = \frac{1}{2\pi}\int_0^{2\pi} U(\xi)\,d\xi = 0 \ . \tag{8.35}$$

The condition (8.35) can be easily satisfied at zero discrepancy $\Delta = 0$. For this case the integration constant is zero in equation (8.34), $C = 0$, and the acoustic velocity is an odd function in time:

$$U = \pm\sqrt{\frac{2M}{\pi\epsilon}}\cos\frac{\xi}{2} \ , \quad -\pi < \xi < \pi \ . \tag{8.36}$$

The "+"-sign is valid at $-\pi < \xi < 0$, and the "-"-sign is valid for the positive half-period $0 < \xi < \pi$. The wave (8.36) has a shock at $\xi = 0$. A simple solution exists also at large discrepancies, $|\Delta| \gg M$, if it is possible to eliminate the nonlinear term; in this case $U = -(M/2\Delta)\cos\xi$. In the general case, for arbitrary values of discrepancy, the solution of the "simple" quadratic equation has a complicated form[8].

One period of the steady-state wave is shown in Figure 8.2 as a function of "fast" time, and for different values of the discrepancy.

---

[8]See details in paper: B.O.Enflo, C.M.Hedberg, O.V.Rudenko, J.Acoust.Soc. Am. **117**(2), 601-612, 2005.

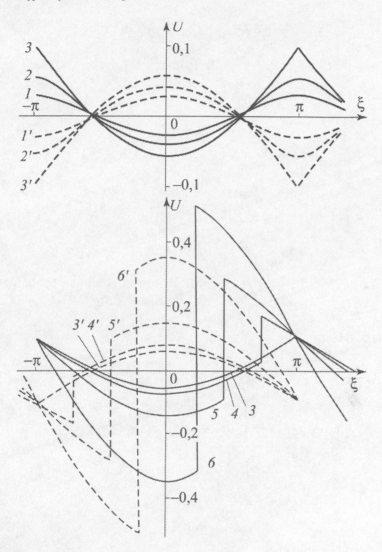

Figure 8.2: *Profiles of one period of standing wave described by the inhomogeneous Riemann equation for waves with discrepancy. The corresponding acoustic Mach numbers increase with increase in the number of the curve 1-6.*

Problem 8.8

Generalize equation (8.27) in Problem 8.5, by introducing dissipation.

Answer

$$\frac{\partial U}{\partial T} + \Delta \frac{\partial U}{\partial \xi} - \pi \epsilon U \frac{\partial U}{\partial \xi} - D \frac{\partial^2 U}{\partial \xi^2} = \frac{M}{2} \sin \xi , \qquad D = \frac{b\omega^2 L}{2c^3 \rho} . \quad (8.37)$$

This equation is known as the 'Inhomogeneous Burgers Equation"
(IBE)[9].

Problem 8.9

Calculate the steady-state solution to the IBE, in Problem 8.8, at zero discrepancy, using the Hopf-Cole substitution (see Problem 3.6).

Solution The steady-state solution to IBE is described by the ordinary differential equation

$$D \frac{d^2 U}{d\xi^2} + \pi \epsilon U \frac{dU}{d\xi} = \frac{M}{2} \sin \xi , \quad (8.38)$$

which follows from (8.37) in Problem 8.8, for $T \to \infty$, $\Delta = 0$. Integrating (8.38) we obtain

$$D \frac{dU}{d\xi} + \frac{\pi \epsilon}{2} (U^2 - C) = -\frac{M}{2} \cos \xi . \quad (8.39)$$

Because the mean value $\overline{U} = 0$, it follows from equation (8.39) that the constant equals

$$C = \overline{U^2} = \frac{1}{2\pi} \int_0^{2\pi} U^2(\xi) \, d\xi . \quad (8.40)$$

So, the constant C has an important physical meaning: it is proportional to the acoustic energy density[3]. Because the acoustic field

---

[9]See paper: O.V.Rudenko. JETP Lett. **20**(7), 203-204, 1974.

inside the resonator is composed by two counter-propagating waves, the energy is twice higher.

Using the Hopf-Cole transformation

$$U = \frac{2D}{\pi\epsilon}\frac{d}{d\xi}\ln W , \qquad (8.41)$$

we reduce the nonlinear equation (8.39) to a linear one for Mathieu functions

$$\frac{d^2W}{dz^2} + [-(\frac{\pi\epsilon}{D})^2 C + \frac{\pi\epsilon M}{D^2}\cos(2z)] = 0 , \qquad (8.42)$$

where $z = \xi/2$. Comparing equations (8.42) and (8.40) we conclude that the energy is proportional to the eigenvalue λ_0 of the Mathieu function ce_0 [22]:

$$\overline{U^2} = -(\frac{D}{\pi\epsilon})^2 \cdot \lambda_0(q = \frac{\pi\epsilon M}{2D^2}) . \qquad (8.43)$$

The total energy of the resonator equals $E = \rho c^2 V \cdot 2\overline{U^2}$, where V is the resonator volume. At weak excitation $\lambda_0 \approx -q^2/2$ [22], a well-known linear result appears :

$$E \approx (\frac{M}{2D})^2 \cdot \rho c^2 V . \qquad (8.44)$$

Using another asymptotic for eigenvalue λ_0 with $q \gg 1$ [22], for a strong boundary vibration we derive

$$E = [\frac{2M}{\pi\epsilon} - \frac{2D}{(\pi\epsilon)^2}\sqrt{2\pi\epsilon M} + \frac{1}{2}\frac{D^2}{(\pi\epsilon)^2} + \ldots] \cdot \rho c^2 V . \qquad (8.45)$$

A principal difference between energy at weak and at strong excitation exists. In the weak linear case the energy (8.44) is proportional to the square of the velocity amplitude of the wall vibration M^2. On the other hand, at strong excitation (8.45) the process of nonlinear absorption of shock waves in the cavity of resonator leads to a slower dependence - proportional to M.

Problem 8.10

Let the light spot of laser beam on a water surface move along the surface in the x-direction with a velocity c which is close to the sound velocity c_0. The heating of water is described by the heat transport equation

$$\rho c_p \frac{\partial T}{\partial t} = \kappa \Delta T + \alpha I_0 \exp(-\alpha z) f(x - ct) . \qquad (8.46)$$

The thermo-elastic stress that appears in the water due to its heating excites a sound wave governed by the inhomogeneous wave equation:

$$\frac{\partial^2 u}{\partial x^2} - \frac{1}{c_0^2} \frac{\partial^2 u}{\partial t^2} + \frac{\epsilon}{c_0^3} \frac{\partial^2 u^2}{\partial t^2} = \beta \frac{\partial^2 T}{\partial x \partial t} . \qquad (8.47)$$

Here β is the coefficient of thermal expansion, α is the absorption coefficient of light in water, c_p is the specific heat of water (at fixed pressure) and I_0 is the intensity of the laser beam on the water surface $z = 0$. The function $f(x)$ describes the cross-section distribution of light intensity of the laser beam - also of the water surface if the beam is immovable. For a fast motion of the light spot when the heat diffusion is negligible, and for weak absorption of light, derive a simplified evolution equation describing excitation and propagation of the sound wave [6, 7].

Solution Put $\kappa = 0$ and $\exp(-\alpha z) \approx 1$ in (8.46) and substitute the result into equation (8.47): .

$$\frac{\partial^2 u}{\partial x^2} - \frac{1}{c_0^2} \frac{\partial^2 u}{\partial t^2} + \frac{\epsilon}{c_0^3} \frac{\partial^2 u^2}{\partial t^2} = \alpha c_0 N \frac{\partial}{\partial x} f(x - ct) . \qquad (8.48)$$

Here $N = \beta I_0 / (c_0 c_p \rho)$ is the dimensionless number characterizing the transformation of light energy to sound wave energy through the fast thermal expansion of the medium. Seek a solution to equation (8.48) of the form

$$u = u(\xi = x - c_0 t, t_1 = \mu t) \qquad (8.49)$$

which is a modification of the slowly varying profile method. As distinct from the standard form (see (1.9) in Problem 1.5), the slow variable is time instead of distance. After the usual simplifications we derive the evolution equation

$$\frac{\partial u}{\partial t} + \epsilon u \frac{\partial u}{\partial \xi} = \frac{\alpha c_0^2}{2} N \cdot f(\xi - \Delta t) , \qquad \Delta = c - c_0 . \qquad (8.50)$$

If we introduce the new variable $\eta = \xi - \Delta t$, equation (8.50) will take the form

$$\frac{\partial u}{\partial t} - \Delta \frac{\partial u}{\partial \eta} + \epsilon u \frac{\partial u}{\partial \eta} = \frac{\alpha c_0^2}{2} N \cdot f(\eta) , \qquad (8.51)$$

which is known as the inhomogeneous Riemann wave equation (IRE), see (8.27) in Problem 8.5.

Problem 8.11

Solve equation (8.50) for the linear case ($\epsilon = 0$) and analyze the wave resonance phenomenon when $c \to c_0$. This means, that the source exciting the acoustic waves is moving with a speed near the sound velocity.

Answer The solution is similar to that derived in Problem 8.6. By using the notations used there, one writes the solution as

$$u = \frac{\alpha c_0^2}{2} N \frac{F(\xi) - F(\xi - \Delta t)}{\Delta} , \qquad f(\xi) = \frac{dF(\xi)}{d\xi} . \qquad (8.52)$$

Wave resonance takes place at zero discrepancy, and then the sound wave demonstrates an infinite growth:

$$u(x,t) = \frac{\alpha c_0^2}{2} N \cdot t \cdot f(x - c_0 t) . \qquad (8.53)$$

The behaviour of the acoustic pulse is illustrated by Figure (8.3a) (at resonance) and Figure (8.3b) (at non-resonant conditions). Wave resonances are known for waves of different physical origin. They can be easily observed for waves propagating along the water surface. Really, in order to excite high waves on the sea surface, the velocity of the

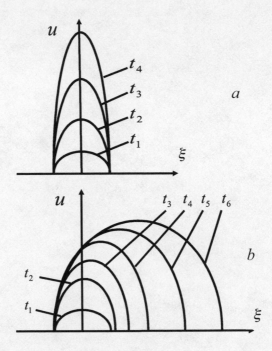

Figure 8.3: *Excitation of an acoustic pulse at wave resonance at zero discrepancy (a); and at non-zero discrepancy (b).*

wind (the driving force) must be approximately equal to the velocity of freely propagating waves. Another example is the sonic boom wave excited by aircraft moving with transsonic velocity. This wave is strongest if the aircraft moves with a velocity somewhat exceeding c_0. However, the resistive force is also the largest at such velocities.

Problem 8.12

Reduce equation (8.51) in Problem 8.10, to its simplest form containing only one independent coefficient.

Solution To simplify the equation, we introduce new dimensionless variables

$$\bar{u} = \frac{u}{u_0} , \qquad \bar{t} = \frac{t}{t_0} , \qquad \bar{\eta} = \frac{\xi - \Delta t}{a} . \tag{8.54}$$

The constant a is already defined, it is the characteristic width of the laser beam. The two other constants are unknown. Using the variables (8.54), one can rewrite the equation (8.51)

$$\frac{u_0}{t_0}\frac{\partial \bar{u}}{\partial \bar{t}} - \Delta \frac{u_0}{a}\frac{\partial \bar{u}}{\partial \bar{\eta}} + \epsilon \frac{u_0^2}{a}\bar{u}\frac{\partial \bar{u}}{\partial \bar{\eta}} = \frac{\alpha c_0^2}{2} N \cdot f(\bar{\eta}) . \tag{8.55}$$

Equation (8.55) will have its simplest form

$$\frac{\partial \bar{u}}{\partial \bar{t}} - \delta\frac{\partial \bar{u}}{\partial \bar{\eta}} + \bar{u}\frac{\partial \bar{u}}{\partial \bar{\eta}} = f(\bar{\eta}) . \tag{8.56}$$

at

$$t_0 = \frac{a}{\Delta} , \qquad u_0 = \left(\frac{a\alpha c_0^2 N}{2\epsilon}\right)^{1/2} . \tag{8.57}$$

The only coefficient in the simplified equation (8.56) is

$$\delta = \frac{M - 1}{\sqrt{0.5\epsilon\,\alpha\,aN}} . \tag{8.58}$$

At subsonic movement of the laser beam, when $M = c/c_0 < 1$, the coefficient (8.58) is negative, but at supersonic motion it is positive. So, a similarity law is established here; namely, all one-dimensional movements of the laser beam (of the given form $f(x/a)$) having equal numbers δ are self-similar and can be transformed from one to another by transformation (8.54), and (8.57). This law is valid also, when a still beam is surrounded by a streaming gas moving with transsonic velocity[10]. An analogous transsonic similarity law by Karman-Falkovich is known in gas-dynamics for streamlining of rigid bodies.

As is shown in Problem 8.11, the amplitude of a weak (linear) acoustic wave tends to infinity if the velocity of the laser beam approaches the sound velocity, $c \to c_0$ or $\Delta \to 0$. The solution of the nonlinear equation (8.56) shows, that this amplitude has a finite maximum value reached at a velocity somewhat shifted to supersonic region [4].

---

[10]See: A.A.Karabutov, O.V.Rudenko. Sov.Phys.Acoust. **25**(4), 306-309, 1979.

Problem 8.13[11]

Derive the evolution equation describing nonlinear wave propagation in a scattering medium where the energy loss is proportional to the fourth power of the frequency and the following dispersion law is valid:

$$k = \frac{\omega}{c} + i\beta\omega^4 . \tag{8.59}$$

Solution By replacing the wave number and frequency by the following differential operators

$$k = \frac{1}{i}\frac{\partial}{\partial x} , \qquad \omega = -\frac{1}{i}\frac{\partial}{\partial t} , \tag{8.60}$$

we derive a linear equation

$$\frac{\partial u}{\partial x} + \frac{1}{c}\frac{\partial u}{\partial t} = -\beta\frac{\partial^4 u}{\partial t^4} . \tag{8.61}$$

Because nonlinearity is weak, the corresponding term can be inserted into equation (8.61) additively, as is done at the derivation of the Burgers equation (see Problem 3.2). The final form of the nonlinear equation for a scattering medium is

$$\frac{\partial u}{\partial x} - \frac{\epsilon}{c^2}u\frac{\partial u}{\partial \tau} = -\beta\frac{\partial^4 u}{\partial \tau^4} . \tag{8.62}$$

The retarded time is $\tau = t - x/c$, as before. So, equation (8.62) is similar to Burgers' equation, but contains the fourth derivative instead of the second.

Using dimensionless variables and constants

$$V = \frac{u}{u_0} , \qquad \theta = \omega\tau , \qquad z = \frac{\epsilon}{c^2}\omega u_0 x , \qquad R = \frac{\beta\omega^3 c^2\rho}{\epsilon u_0} , \tag{8.63}$$

one can reduce the equation (8.62) to the simpler form

$$\frac{\partial V}{\partial z} - V\frac{\partial V}{\partial \theta} = -R\frac{\partial^4 V}{\partial \theta^4} . \tag{8.64}$$

---

[11]Problems 8.13 and 8.14 are based on results of the paper: O.V.Rudenko, V.A.Robsman, Doklady Physics (Reports of Russian Academy of sciences) **47**(6), 443-446, 2002

Problem 8.14

Show that an exact steady-state solution to equation (8.64) in problem 8.13 is given by the implicit integral formula

$$\theta = (\frac{40}{9}R)^{1/3} \cdot \int_0^V \frac{dy}{(1-y^2)^{2/3}} \,. \tag{8.65}$$

Show that this equation *cannot* have a steady-state solution like the shock wave described by Burgers equation

$$V = \tanh[\frac{\theta}{(16R)^{1/3}}] \,. \tag{8.66}$$

Solution One can easily check that solution (8.65) satisfies the ordinary differential equation

$$V\frac{dV}{d\theta} = R\frac{d^4V}{d\theta^4} \tag{8.67}$$

– it follows from equation (8.64) at $\theta \to \infty$. The particular solution (8.65) is an odd function; it increases monotonically from $V(0) = 0$ and attains the value $V(\theta_*) = 1$ in a finite time $\theta = \theta_*$. The monotonic solution (8.66), which tends to unity at $\theta \to \infty$, satisfies another equation

$$(V - \frac{3}{4}V^3)\frac{dV}{d\theta} = R\frac{d^4V}{d\theta^4} \,, \tag{8.68}$$

which differs from (8.66) in its type of nonlinearity. The form of the shock which tends asymptotically to unity, $V(\theta) \to 1$, and satisfies equation (8.67), has a non-monotonic behaviour. It has been numerically shown that the shock front contains decaying oscillations caused by strong absorption of higher harmonics.

Problem 8.15

In a relaxing medium the deviation of density ρ' caused by an acoustic wave depends on the acoustic pressure p' through the integral relation

$$c_0^2\rho' = p' - m \int_{-\infty}^t \exp(-\frac{t-t'}{t_{rel}})\frac{dp'}{dt'} \, dt' \,. \tag{8.69}$$

The integral indicates that the density $\rho'(t)$ at a given moment depends not only on the pressure measured at the same moment t, but on the history of the function $p'(t)$. In other words, the density depends on the behaviour of the pressure of the past, from $t = -\infty$ to the present time. Of course, the "memory" of the medium relaxes and the distant past has a weaker influence. According to formula (8.69), the memory decays exponentially, as $\exp(-t/t_{rel})$, where t_{rel} is the characteristic relaxation time. Equation (8.69) is referred to as the determining equation, used instead of the algebraic equation of state $p'(\rho')$ used before. Let the pressure increase as

$$p'(t) = P_0 H(t) , \tag{8.70}$$

where $H(t)$ is the unit step function: $H(t) = 0$, $t < 0$; $H(t) = 1$, $t > 0$. Calculate the corresponding step for the medium density $\rho'(t)$.

Answer Substituting (8.70) into the determining equation (8.69), we calculate

$$c_0^2 \rho' = P_0[1 - m \exp(-t/t_{rel})] . \tag{8.71}$$

The density (8.71) jumps at the moment $t = 0$ from zero to $\rho' = (P_0/c_0^2)(1 - m)$ and then approaches asymptotically its steady-state value $\rho' = (P_0/c_0^2)$. At the same time, the pressure instantly reaches its steady-state $p'(t) = P_0$. The delay of the density is caused by ongoing internal processes in the medium with characteristic relaxation time t_{rel}. The number m characterizes the strength of the relaxation process - usually this number is small.

Problem 8.16

Using linear hydrodynamic equations and the determining equation (8.69) in Problem 8.15, derive a linear wave equation for plane waves propagating in a relaxing medium. Calculate the frequency-dependent velocity of propagation and the absorption coefficient for this wave.

Solution Eliminating the particle velocity from the equation of continuity and the equation of motion, we derive

$$\frac{\partial^2 p'}{\partial x^2} = \frac{\partial^2 \rho'}{\partial t^2} .\qquad(8.72)$$

Now it is necessary to substitute the determining equation into (8.72) and get

$$\frac{\partial^2 p'}{\partial x^2} - \frac{1}{c_0^2}\frac{\partial^2 p'}{\partial t^2} = -\frac{m}{c_0^2}\frac{\partial^2}{\partial t^2}\int_{-\infty}^{t}\exp(-\frac{t-t'}{t_{rel}})\frac{dp'}{dt'}\,dt' .\qquad(8.73)$$

To obtain the dispersion law we seek for the solution to (8.73) in form of a monochromatic wave:

$$p' = P_0\exp(-i\omega t + ikx) .\qquad(8.74)$$

It follows from (8.73) and (8.74) that the wave number depends on frequency as

$$k^2 = \frac{\omega^2}{c_0^2}[1 + im\frac{\omega t_{rel}}{1 - i\omega t_{rel}}] .\qquad(8.75)$$

The real and imaginary parts of the wave number for the wave traveling in positive direction along the x-axis are

$$k = k' + ik'' ,\qquad k' \approx \frac{\omega}{c_0}[1 - \frac{m}{2}\frac{\omega^2 t_{rel}^2}{1 + \omega^2 t_{rel}^2}] ,\qquad k'' \approx \frac{m}{2c_0 t_{rel}}\frac{\omega^2 t_{rel}^2}{1 + \omega^2 t_{rel}^2} .$$
$$(8.76)$$

The smallness of m is taken into account at the simplification in (8.76). From this formula we derive results for propagation velocity and absorption coefficient:

$$c(\omega) = c_0(1 + \frac{m}{2}\frac{\omega^2 t_{rel}^2}{1 + \omega^2 t_{rel}^2}) ,\qquad \alpha = \frac{m}{2c_0 t_{rel}}\frac{\omega^2 t_{rel}^2}{1 + \omega^2 t_{rel}^2} .\qquad(8.77)$$

One can see that the velocity of propagation increases with increase in frequency. The absorption at the path length equal to the wave length has maximum for $\omega t_{rel} = 1$. The following relations exist:

$$c(\omega = 0) = c_0 ,\qquad c(\omega \to \infty) \to c_0(1 + m/2) ,$$

$$\alpha\lambda = \pi m\frac{\omega t_{rel}}{1 + \omega^2 t_{rel}^2} ,\qquad (\alpha\lambda)_{max} = \pi\frac{m}{2} .\qquad(8.78)$$

Problem 8.17

Derive the evolution equation describing nonlinear wave propagation in a relaxing medium. Write the simplified equation using normalized variables like (8.63) in Problem 8.13.

Solution The evolution equation is derived by a standard slowly varying profile method. Because nonlinearity is weak, the corresponding term can be inserted into the equation additively, as is done at the derivation of Burgers' equation (see Problem 3.2). The final form of equation is

$$\frac{\partial p'}{\partial x} - \frac{\epsilon}{c_0^3 \rho_0} p' \frac{\partial p'}{\partial \tau} = \frac{m}{2c_0} \frac{\partial}{\partial \tau} \int_{-\infty}^{\tau} \exp(-\frac{\tau - \tau'}{t_{rel}}) \frac{\partial p'}{\partial \tau'} \, d\tau' \ . \tag{8.79}$$

The exponential kernel in the integral term of the equation (8.79) makes it possible to write it in the differential form:

$$(1 + t_{rel} \frac{\partial}{\partial \tau}) [\frac{\partial p'}{\partial x} - \frac{\epsilon}{c_0^3 \rho_0} p' \frac{\partial p'}{\partial \tau}] = \frac{m t_{rel}}{2c_0} \frac{\partial^2 p'}{\partial \tau^2} \ . \tag{8.80}$$

Problem 8.18

Derive the steady-state solution to equation (8.80) in Problem 8.17 describing a shock wave in a relaxing medium.

Solution The steady-state solution does not depend on distance, and the derivative is zero. It means that the wave keeps its shape constant. For a stationary wave the partial differential equation (8.80) transforms itself to an ordinary differential equation

$$(1 + t_{rel} \frac{d}{d\tau}) p' \frac{dp'}{d\tau} + P_0 t_{rel} D \frac{d^2 p'}{d\tau^2} = 0 \ , \qquad D = \frac{m c_0^2 \rho_0}{2\epsilon P_0} \ . \tag{8.81}$$

Integrating (8.81) once with account for the boundary condition at $\tau \to \infty$

$$p' = P_0 \ , \qquad \frac{dp'}{d\tau} = 0 \ . \tag{8.82}$$

we obtain

$$\frac{dp'}{d\tau} = \frac{1}{2P_0 t_{rel}} \frac{P_0^2 - p'^2}{D + p'} . \tag{8.83}$$

The second integration leads to an algebraic solution describing a stationary wave (see [4]):

$$\frac{\tau + C}{t_{rel}} = \ln \frac{(1 + p'/P_0)^{D-1}}{(1 - p'/P_0)^{D+1}} . \tag{8.84}$$

At large values of the parameter $D \gg 1$, corresponding to a weak nonlinear effect, the solution (8.84) reduces to

$$p' = P_0 \tanh(\frac{\tau + C}{2D t_{rel}}) . \tag{8.85}$$

Formula (8.85) describes a symmetric shock wave of the same usual form as in a dissipative medium. At decreasing D, when $D > 1$ the jump becomes non-symmetric; the tail section of the shock front is delayed by a relaxation process. In the case of strongly expressed nonlinearity, $D < 1$, the formula (8.84) describes a two-valued function having no physical meaning. Consequently, the smoothening by relaxation cannot compete with the nonlinear steepening, and dissipation must be added. In other words, equation (8.72) must be complemented by the term with the second derivative which is found in the Burgers equation.

REFERENCES

1. Physical Acoustics, Mason W.P., (Ed.) Vol.II Part B.
    Academic Press, 1965.
2. High Intensity Ultrasonic Fields, Rosenberg L.D. (Ed.)
    Plenum Press, 1974.
3. Beyer R.T. Nonlinear Acoustics.
    Naval Ship Systems Command, 1974
    Am.Inst.of Physics, 1997 (2nd edition).
4. Rudenko O.V., Soluyan S.I. Theoretical Foundations
    of Nonlinear Acoustics. Plenum Press, 1977.
5. Whitham G.B. Linear and Nonlinear Waves. J.Wiley & Sons, 197
6. Vinogradova M.B., Rudenko O.V., Sukhorukov A.R. Theory of W
    Moscow, Nauka, 1979 (1st Edition); 1990 (2nd Edition) (in Russia
7. Novikov B.K., Rudenko O.V., Timoshenko V.I. Nonlinear
    Underwater Acoustics. American Inst.of Physics, 1987.
8. Guyer R.A., Johnson P.A., Nonlinear mesoscopic
    elasticity: evidence for a new class of materials
    Physics Today April 1999, 30-36.
9. Bailey M.R., Khoklova V.A., Sapozhnikov O.A., Kargl S.G.,
    Crum L.A., Physical mechanisms of the therapeutic effect of ultra
    Acoustical Physics **49**(4), 369-388, 2003.
10. Vasileva ,O.A., Karabutov O.A., Lapshin E.A., Rudenko O.V.
    Interaction of one-dimensional waves in media without dispersion
    Moscow University Press, 1983 (in Russian).
11. Bakhvalov N.S., Zhileikin Ya.M., Zabolotskaya E.A.
    Nonlinear Theory of Sound Beams.
    American Inst. of Physics, 1987.
12. Engelbrecht Y.K., Fridman V.E., Pelinovskii E.N.
    Nonlinear Evolution equations
    Pitman Res. Notes Math. **180**, Longman, London 1988.
13. Gurbatov S.N., Malakhov A.N., Saichev A.I.
    Nonlinear Random Waves and Turbulence in Nondispersive
    Media. Waves, Rays, Particles
    Manchester University Press, 1991.

14. Rudenko O.V. Interactions of Intense Noise Waves
    Sov.Phys. Usp. **29**(7), 620-64, 1986.
15. Gurbatov S.N., Saichev A.I., Jakushkin I.G.
    Nonlinear Waves and One-Dimensional Turbulence in
    Nondispersive Media
    Sov.Phys.Usp. **26**(10), 857-876, 1983.
16. Enflo B.O, Hedberg C.M.
    Theory of nonlinear acoustics in fluids
    Kluwer Academic Publishers, Dordrecht, 2002.
17. Nonlinear Acoustics, Hamilton M.F., Blackstock D.T. (Eds)
    Academic Presss, San Diego, 1997.
18. Naugolnykh K.A, Ostrovsky L.A.,
    Nonlinear wave processes in acoustics
    Cambridge University Press, New York, 1998.
19. Gusev V.E., Karabutov A.A., Laser Optoacoustics
    American Institute of Physics, New York, 1993.
20. Rudenko O.V., Sapozhnikov O.A.
    Self-action effects for wave beams containing shock fronts
    Physics-Uspekhi **47**(9), 907-922, 2004.
21. Rudenko O.V, Nonlinear sawtooth-shaped waves
    Physics-Uspekhi **38**(9), 965-990, 1995.
22. Handbook of Mathematical Functions
    Abramowitz M. and Stegun I.A.,(Eds)
    National Bureau of Standards, 1964.

SUBJECT INDEX